560-177

D1420861

160564

Nutritional Qualities of Fresh Fruits and Vegetables

Edited By

Philip L. White, Sc. D.
Secretary, Council on Foods and Nutrition and
Director, Department of Foods and Nutrition
American Medical Association

and

Nancy Selvey, R. D.
Nutritionist, Department of Foods and Nutrition
American Medical Association

TX557
N87

FUTURA
PUBLISHING COMPANY
1974

Copyright © 1974
Futura Publishing Company, Inc.

Published by
Futura Publishing Company, Inc.
295 Main Street
Mount Kisco, New York 10549

L.C.: 73-89312
ISBN: 0-87993-036-5

All rights reserved.
No part of this book may be translated or reproduced in any form without the written permission of the publisher.

Printed in U.S.A. by
NQBLE OFFSET PRINTERS, INC.
New York, N.Y. 10003

12.50

Foreword

This book consists of papers on the nutritional qualities of fresh fruits and vegetables presented at a symposium held November 9 and 10, 1972.* It evaluates present knowledge of the roles played by fruits and vegetables in the nutritional quality of American diets and explores the influence of certain pre- and post-harvest practices on the aesthetic and nutritional qualities of fresh products.

Specialists in plant breeding, agronomy, horticulture, plant pathology, transportation, storage, food science, marketing, food analysis, and nutrition, discussed advances in agriculture and technology and worked together to define certain pressing problems and to propose priorities for their resolution.

The innovations in agriculture, transportation and marketing that have provided fresh produce the year around are more or less taken for granted. These great advances in a scant generation have not eliminated the need for vigilance throughout the food chain because at each step along the way there lurks the possibility for nutritional and aesthetic damage to the products. In order to provide fresh fruits and vegetables throughout the year they must be stored, transported often under adverse conditions over great distances, perhaps again stored, and finally merchandised. Each step potentially may result in some losses in nutritional and other qualities.

Are there alternatives to a chain with so many links of precarious strength? The approach of this book is to review information available on each link in the chain from farm to table evaluating where improvements can be made. Until post-harvest handling can be truly perfected so as to assure optimum retention of nutrients, is it worthwhile to attempt to increase nutrient values through plant breeding techniques? Or is it

* Sponsored by the AMA Council on Foods and Nutrition with the Cooperation of the USDA.

enough to hold present nutrient levels while breeding for such needs as increased yields and the ability to withstand mechanical harvesting and bulk handling? An overriding requirement for breeders is to produce plants that are pest-resistant since present federal orders demand reduced dependence on pesticides. Another obligation is proper genetic selection and agricultural management to avoid or suppress undesirable qualities. Are not these needs of greater priority than that of increasing nutritive values? If not, which nutrients ought to be given priority?

These are important questions since genetic research may require 10-15 or more years before the mission is accomplished. The nutritionist, if he is to influence the acceptance of priorities, must become involved at an early stage in decision making.

Much of the production of fruits and vegetables is now concentrated in our far Southwest. This is the result of a trend to locate production in areas of maximum yields. How does this influence composition and aesthetic qualities? Do the increased handling and transportation needs so imposed justify such concentrations of crops? How do we integrate the convenience for processing and the inconvenience of transportation precipitated by centralization of agriculture? Centralization clearly places great demands upon our transportation system; much needs to be done to improve this system so as to take advantage of longer growing periods possible in the Southwest.

These, then, are a few of the decisions that must be made as we struggle to retain, nay, improve the place of fresh fruits and vegetables in the diets of Americans.

Throughout this book one finds many references to the ascorbic acid content of fresh products. This emphasis on ascorbic acid is quite excusable on the grounds that it is easily determined analytically and, being a rather fragile compound, is a very convenient benchmark of the retention or loss of nutrients as the product is handled and stored. Additionally, since about 95% of dietary ascorbic acid is obtained from fruits and vegetables, it is well that some consideration is given to its retention. Vitamin A and its precursors are also given great attention, not only as they relate to color characterizations, but also because fruits and vegetables provide about 50% of this important vitamin. There is little attention given to the protein content of fresh products. This slighting of protein is in no way meant to depreciate its value or the role of the plant kingdom in providing protein for much of the world's population. Rather, contributors were requested to concentrate their attention on vitamins and minerals. To achieve continuity, some reservations were called for.

An added feature of this book is the inclusion of reports of task forces assembled during the November 1972 meeting. The task forces, consisting of participants and speakers, considered priorities for research and education relating to the production and utilization of fresh fruits and vegetables.

A feeling common to all groups was that of a need for more cooperation among individuals concerned with genetics, agriculture, food science and nutrition. It is recommended that several interdisciplinary research institutes housing these specialities be established. The problems emerging as farming practices change, population increases, and food habits evolve require new attacks on the old problem of poor communication. It rather boils down to the need for those with special skills and laboratories to cooperate more fully with the geneticist and horticulturalist. Nutrient analyses of new strains would be routine for food analysts but probably foreign to the geneticist. In order that the food scientist and nutritionist may influence the composition of new strains, these persons must be readily available for consultation and laboratory support.

The production of food crops of the highest nutritional values which can withstand the rigors of post-harvest handling so as to arrive in the larders of consumers at the peak of goodness is the goal. To this end, this book is dedicated.

Philip L. White, Sc.D.

Contributors

Anderson, Dale L., Ph.D., Staff Scientist for Transportation and Facilities, National Program Staff, Agricultural Research Service, U.S. Department of Agriculture, Hyattsville, Maryland

Ang, Catharina Y.W., Ph.D., Laboratory Manager, Food Science Associates, Inc. Dobbs Ferry, New York

Chichester, C.O., Ph.D., Professor of Food and Resource Chemistry, University of Rhode Island, Kingston, Rhode Island

Crosby, Edwin A., Ph.D., Director, Agriculture Division, National Canners Association, Washington, D. C.

Gabelman, W.H.,, Prof., Chairman, Department of Horticulture, University of Wisconsin, Madison, Wisconsin

Gebhardt, Susan E., Nutrient Data Research Center, Consumer and Food Economics Institute, Agricultural Research Service, U.S. Department of Agriculture, Hyattsville, Maryland

Heinze, P.H., Ph.D., Staff Scientist, National Program Staff, Agricultural Research Service, U.S. Department of Agriculture, Beltsville, Maryland

Kehr, August E., Ph.D., Staff Scientist, Plant and Entomological Sciences, National Program Staff, Agricultural Research Service, U.S. Department of Agriculture, Beltsville, Maryland

Lee, T.C., Ph.D., Department of Food and Resource Chemistry, University of Rhode Island, Kingston, Rhode Island

Livingston, G.E., Ph.D., Food Science Associates, Inc., Dobbs Ferry, New York

Maier, V.P., Ph.D., Laboratory Chief, Fruit and Vegetable Chemistry Laboratory, Western Region, Agricultural Research Service, U.S. Department of Agriculture, Pasadena, California

Meinken, Margaret, R.D., Food Analyst, Central Food Stores, University of Missouri, Columbia, Missouri

Murphy, Elizabeth W., S.M., Research Chemist, Nutrient Data Research Center, Consumer and Food Economics Institute, Agricultural Research Service, U.S. Department of Agriculture, Hyattsville, Maryland

Peterkin, Betty, Consumer and Food Economics Institute, Agricultural Research Service, U.S. Department of Agriculture, Hyattsville, Maryland

Rizek, Robert L., Ph.D., Chairman, Consumer and Food Economics Institute, Agricultural Research Service, U.S. Department of Agriculture, Hyattsville, Maryland

Salomon, Milton, Ph.D., Prof., Department of Food and Resource Chemistry, University of Rhode Island, Kingston, Rhode Island

Stevens, M. Allen, Assistant Geneticist, Department of Vegetable Crops, University of California, Davis, California

Swope, Daniel A., Consumer and Food Economics Institute, Agricultural Research Service, U.S. Department of Agriculture, Hyattsville, Maryland

Watt, Bernice K., Ph.D., Nutrition Analyst, Nutrient Data Research Center, Consumer and Food Economics Institute, Agricultural Research Service, U.S. Department of Agriculture, Hyattsville, Maryland

Yokoyama, H., Fruit and Vegetable Chemistry Laboratory, Western Region, Agricultural Research Service, U.S. Department of Agriculture, Pasadena, California

Table of Contents

Trends in Fresh Fruit and Vegetable Consumption and Their Nutritional Implications*

Robert L. Rizek, Ph.D., Daniel A. Swope and Betty Peterkin

Most people, regardless of their nutritional knowledge, look upon fruits and vegetables as important foods for good health. But whether they support this belief by actually selecting these foods is another question. A study of the trends in consumption of fruits and vegetables might provide some answers. Before reviewing these trends, however, we should gain a perspective of the importance of fruits and vegetables, as a food group, in providing certain nutrients in the nation's food supply.

NUTRIENTS CONTRIBUTED BY FRUITS AND VEGETABLES

In 1971, almost all of the ascorbic acid (over 90%) in the food supply was provided by fruits and vegetables (Fig. 1). Furthermore, one-half of the vitamin A value, about one-third of the vitamin B_6, and just about one-fifth of the thiamin and niacin were provided by this food group. Among the minerals, about one-fourth of the magnesium and one-fifth of the iron were provided by fruits and vegetables. Fresh fruits and vegetables alone contributed over one-half of the ascorbic acid, about one-third of the vitamin A, one-fourth of the vitamin B_6, one-sixth of the magnesium, one-eighth of the thiamin and one-

---

* Official information material, U.S. Government. This chapter may be reproduced in part or in full.

tenth of the iron. We can safely say that fruits and vegetables as a group are indeed "highly nutritious."

In these days of concern about weight control and fat in the diet, it is of interest to note that less than one-tenth of the calories and only about one percent of the fat in the food supply come from fruits and vegetables.

The nutritional importance of fruits and vegetables in the American diet is dependent upon two principal factors—their nutrient composition and the quantity eaten. Certain vegetables and fruits are recognized for their contribution of certain nutrients. Citrus fruits and fresh strawberries, for example, are rich in ascorbic acid. Other important sources include tomatoes, cabbages, melons, dark green vegetables and potatoes and sweet potatoes, especially when cooked in their jackets. Carotene, which the body can change to vitamin A, is found in large amounts in dark green vegetables, such as kale, spinach, and broccoli, and in deep yellow vegetables—carrots, sweet potatoes and some squash—and in deep yellow fruits such as apricots.

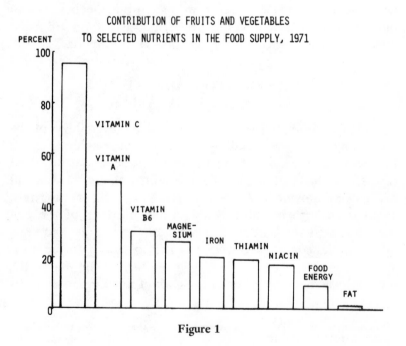

Figure 1

Tomatoes are also important as a source of vitamin A. (See Chapter 2 for information on reassessment of vitamin A values of foods.) Dark green vegetables provide worthwhile amounts of the minerals, iron and magnesium. Dark greens and potatoes are good sources of vitamin B_6.

Nutrient content varies not only with the kind of fruit or vegetable but also with variety and growing conditions. Fresh fruits and vegetables, as well as other foods, lose some nutritive value in the stages between the farm and the table. The amount varies considerably with the nutrient and the food in question and the way the food is handled in the marketing system, and the way it is handled in the home. Important factors affecting nutrient loss in a fresh fruit or vegetable between harvest and use on the table include the length of the time it is stored, temperature and humidity during storage, the amount of physical injury due to handling, and the method of preparation for serving. For example, fresh vegetables, such as kale, spinach, turnip greens, chard, broccoli, and salad greens, need to be chilled or refrigerated as soon as possible after harvest. They keep their nutrients best when held near freezing and at high humidity. Leafy, dark-green vegetables and broccoli keep practically all of their ascorbic acid for several days if they are packed in crushed ice. Many deep-green vegetables have such high initial values that they remain excellent sources of ascorbic acid even after a substantial loss. They could be expected to provide more vitamin C than freshly harvested snap beans and head lettuce— perhaps more even than tomatoes.

Fresh strawberries are such a good source of ascorbic acid that a handful direct from the patch would supply a man his entire day's need of vitamin C. Berries generally are highly perishable and lose much of their ascorbic acid quickly if capped or stemmed or if their tissue becomes bruised. Berries to be held a few days should be kept cold, dry, and whole to retain their maximum values.

Many of the fruits and vegetables we eat are purchased in a processed form—either canned, frozen or dried. Processing has made an important contribution to the variety, quality, and safety of the fruits and vegetables available to us. However, some nutritive value is lost during processing and during the storage of the processed food. Most susceptible to these losses are vitamins that are water soluble and destroyed by heat, such as ascorbic acid and thiamin.

Shown in Table 1 are representative nutritive values for a few selected foods from *Agriculture Handbook No. 8* which indicate the effects of shifting from use of fresh produce to processed forms of fruits or vegetables.

SHIFTS IN SOURCES OF NUTRIENTS

Technological changes in conjunction with shifts in consumer preference for certain fresh fruits and vegetables and for those that are processed have resulted in significant changes in the food sources of the various nutrients over the past 40 to 50 years.

Fruits and vegetables provided about half of the vitamin A in our diets in 1971, about the same as in 1925-29 (Fig. 2). The proportion supplied by sweet potatoes, however, has declined sharply since the 1920's but they still provide about 5% of the total. This decline was offset by an increase in the contribution of all other fruits and vegetables—from 36% in 1925-29 to around 45% in 1971.

About the same proportion of ascorbic acid was obtained from the fruit and vegetable group in 1971 as in the 1925-29 base period, but nearly three times as much is now contributed by citrus fruits as in the 1920's and considerably less by potatoes (Fig. 3). The contribution from tomatoes increased slightly, but that from other vegetables decreased from nearly 30% of the total to just under 20%. The decrease resulted partly from a decline in the use of dark-green leafy vegetables.

The vegetable and fruit group is relatively less important in supplying thiamin than are animal products and grain products (Fig. 4). Over the years, since 1925-29, the contribution of vegetables and fruits, including potatoes, decreased slightly from about 25% of the total thiamin supply to just about 20%. The contribution has remained at about 20% over the past 25 years.

Most of the vitamin B_6 in our diets is obtained from foods of

Table 1.—Composition of Selected Fruits and Vegetables, 100 Grams Edible Portion[1]

Food description and AH-8 Item No.	Iron Mg.	Vitamin A value I.U.	Thiamin Mg.	Ribo-flavin Mg.	Ascorbic acid Mg.	Sodium Mg.	Potassium Mg.
Apricots							
30 Raw	.5	2,700	.03	.04	10	1	281
32 Canned	.3	1,830	.02	.02	4	1	246
Asparagus							
46 Raw spears	1.0	900	.18	.20	33	2	278
47 Cooked spears, boiled and drained	.6	900	.16	.18	26	1	183
49 Green, canned spears, regular pack, drained solids	1.9	800	.06	.10	15	236	166
Lima beans							
164 Raw	2.8	290	.24	.12	29	2	650
165 Cooked, boiled, drained	2.5	280	.18	.10	17	1	422
167 Canned, drained solids	2.4	130	.03	.05	6	236	222
173 Frozen, cooked, boiled, drained	1.7	230	.07	.05	17	101	426
Green snap beans							
182 Raw	.8	600	.08	.11	19	7	243
183 Cooked, boiled and drained	.6	540	.07	.09	12	4	151
186 Canned, drained solids	1.5	470	.03	.05	4	236	95
192 Frozen, cooked, boiled, drained	.7	580	.07	.09	5	1	152
Cauliflower							
630 Raw	1.1	60	.11	.10	78	13	295
631 Cooked, boiled, drained	.7	60	.09	.08	55	9	206
633 Frozen, cooked, boiled, drained	.5	30	.04	.05	41	10	207
Cherries							
662 Raw, sour, red	.4	1,000	.05	.06	10	2	191
665 Canned, sour, red, water pack	.3	680	.03	.02	5	2	130

Table 1 (continued)

Food description and AH-8 Item No.	Iron Mg.	Vitamin A value I.U.	Thiamin Mg.	Riboflavin Mg.	Ascorbic acid Mg.	Sodium Mg.	Potassium Mg.
Sweet corn							
844 Raw	.7	400	.15	.12	12	Trace	280
845 Cooked, boiled, drained, cut off cob	.6	400	.11	.10	7	Trace	165
846 Canned, cream style	.6	330	.03	.05	5	Trace	196
857 Frozen, cooked, boiled, drained	.8	350	.09	.06	5	1	184
Grapefruit							
1059 Raw, all varieties	.4	20	.04	.02	38	1	135
1069 Canned, water pack	.3	10	.03	.02	30	4	144
Peaches							
1479 Raw	.5	1,330	.02	.05	7	1	202
1480 Canned, water pack	.3	450	.01	.03	3	2	137
Peas							
1515 Raw	1.9	640	.35	.14	27	2	316
1516 Cooked, boiled, drained	1.8	540	.28	.11	20	1	196
1518 Canned, regular pack, drained solids	1.9	690	.09	.06	8	236	96
1530 Frozen, cooked boiled, drained	1.9	600	.27	.09	13	115	135
Potatoes							
1785 Raw	.6	Trace	.10	.04	20	3	407
1789 French fried from raw	1.3	Trace	.13	.08	21	6	853
1793 Mashed from raw	.4	170	.08	.05	9	331	331
1800 Dehydrated, mashed, prepared	.5	110	.04	.05	3	290	290
1806 Frozen French fries, heated	1.8	Trace	.14	.02	21	4	652

1/ The data shown here provide an indication of what may be expected when processed forms are substituted for fresh on an equal weight basis. It is not a measure of the effect of processing on nutritive values. Some differences may be attributed to a difference in the selection of varieties used for the fresh market and for the processed product. In the case of fruit items, the dilution with sirup in the canned product also contributes to the difference between the values for the fresh and canned products.

SOURCES OF VITAMIN A

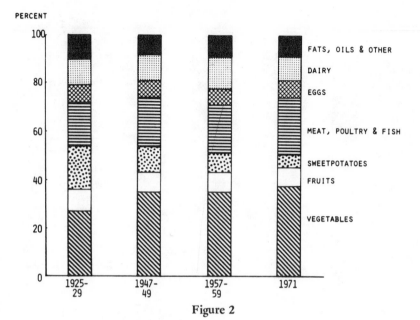

Figure 2

SOURCES OF ASCORBIC ACID

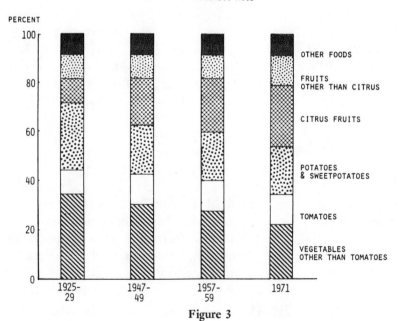

Figure 3

SOURCES OF THIAMIN

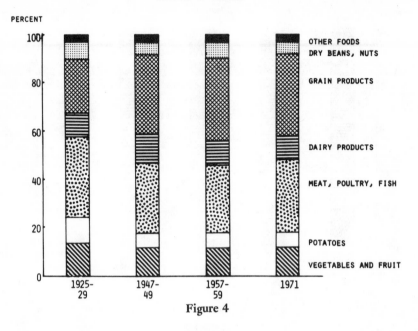

Figure 4

animal origin and they have been providing a greater percentage of this nutrient over the years (Fig. 5). The greatest change since the 1925-29 base period has been in the more than doubling of the proportion coming from the poultry and fish group. On the other hand, vegetables and fruits provided only 30% of the vitamin B_6 in 1971 as against 40% in the 1920's, partly because of declining use of dark-green leaves and potatoes.

Changes in the sources of iron present a pattern somewhat similar to those we have seen for vitamin B_6. More of it has been obtained from the meat and poultry group, which now accounts for about one-third of the total amount (Fig. 6). Vegetables and fruits, including dry beans, dry peas, white and sweet potatoes, have provided a decreasing proportion of the iron in the food supply over the years, from nearly one-third of the total in 1925-29 to around one-fourth of it in 1971.

Because of the decline in consumption of grain products, the magnesium obtained from that food group has decreased rather

SOURCES OF VITAMIN B₆

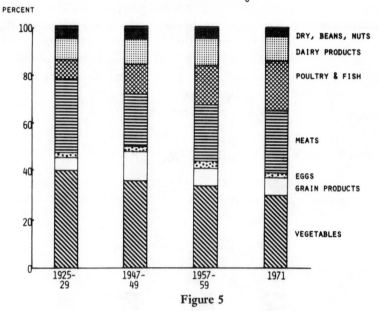

Figure 5

SOURCES OF IRON

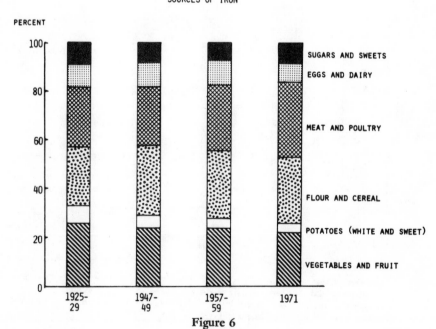

Figure 6

sharply, from about 30% of the total supply in 1925-29 to less than 20% in 1971 (Fig. 7). An increasing amount of the magnesium in the food supply has been obtained from the meat, poultry, and fish group—nearly double the proportions supplied in the base period. Vegetables and fruits, including potatoes, now exceed grain products in supplying magnesium, but the proportion supplied by the vegetable-fruit group has remained remarkably constant for the past 50 years.

FRUIT AND VEGETABLE CONSUMPTION TRENDS*

It is not uncommon to hear that people are eating less vegetables and fruits nowadays and are using more meat, snack

SOURCES OF MAGNESIUM

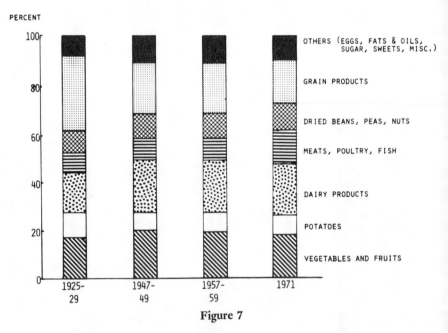

Figure 7

* Unless otherwise specified, all per capita consumption data cited are fresh equivalent of processed plus farm weight used fresh, excluding home production. Statistical Bulletin No. 138 and succeeding supplements "Food Consumption, Prices & Expenditures" Economic Research Service, USDA.

foods, and beverages. As for fruit and vegetable consumption, the direction of the usage trend depends upon how we define fruits and vegetables. If only the fresh form is considered, it is indeed true that there has been a very significant decline over the past four or five decades from an average of 414 pounds per person per year in 1925-29 to 239 pounds in 1971 (Table 2 and Fig. 8). These figures include potatoes and sweet potatoes, which declined from 168 pounds in 1925-29 to only 62 pounds in 1971. On the other hand, there has been a sharp increase in the use of processed fruits and vegetables—from 84 pounds in 1925-29 to 293 pounds in 1971. These weights are the fresh equivalent of processed items.

Table 2.—Yearly Per Capita Consumption of Fruits and Vegetables[1] 1925-29 to 1971

	1925-29	1947-49	1957-59	1971
	Pounds	Pounds	Pounds	Pounds
Fruits				
Citrus				
Fresh	32.4	54.8	34.0	29.3
Processed	.4	35.1	49.0	68.1
Total	32.8	89.9	83.0	97.4
Other				
Fresh	108.5	77.9	61.4	50.8
Processed	38.0	45.2	50.1	48.7
Total	146.5	123.1	111.5	99.5
All fruit				
Fresh	140.9	132.8	95.5	80.1
Processed	38.4	80.3	99.1	116.8
Total	179.3	213.1	194.6	196.9
Vegetables				
Potatoes (incl. sweet)				
Fresh	167.8	125.8	96.5	61.7
Processed	.2	.9	18.7	62.6
Total	168.0	126.7	115.2	124.3
Other				
Fresh	105.0	120.5	104.3	97.3
Processed	45.5	79.2	97.1	113.4
Total	150.5	199.7	201.4	210.7
All vegetables				
Fresh	272.8	246.3	200.8	159.0
Processed	45.7	80.1	115.8	176.0
Total	318.5	326.4	316.6	335.0
All fruits and vegetables				
Fresh	413.7	379.0	296.3	239.1
Processed	84.1	160.3	214.9	292.8
Total	497.8	539.3	511.2	531.9

[1] Fresh equivalent of processed plus farm weight used fresh, excluding home garden produce.

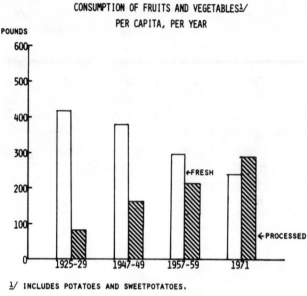

CONSUMPTION OF FRUITS AND VEGETABLES¹/
PER CAPITA, PER YEAR

1/ INCLUDES POTATOES AND SWEETPOTATOES.

Figure 8

   Total consumption of fruits and vegetables, including potatoes, both fresh and processed, increased from 498 pounds per
capita in 1925-29 to 539 pounds in 1947-49. Consumption
declined slightly in the 1950's and early 1960's but has increased in recent years and in 1971 was at about the same level
as in 1947-49.

   Considering only vegetables, other than potatoes, per capita
consumption of fresh vegetables increased from 105 pounds in
the 1920's to over 120 pounds in the 1940's (Fig. 9). Since
then, consumption has trended downward and in 1971, averaged 97 pounds per capita. Consumption of processed vegetables, however, has more than doubled during this period—from
45 pounds in the 1920's to over 113 pounds in the early 1970's.
The total use of vegetables, both fresh and processed increased
from around 150 pounds in the 1920's to over 210 pounds in
1971. In the past two decades, however, per capita consumption of all vegetables has been relatively stable—ranging from a
low of 196 pounds in 1954 to a high of about 214 pounds in
1970.

VEGETABLE CONSUMPTION PER CAPITA[1/]

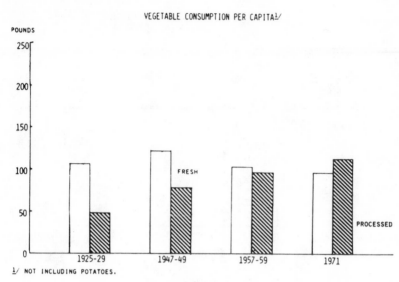

1/ NOT INCLUDING POTATOES.

Figure 9

Consumption of fresh potatoes including sweets declined from 168 pounds in 1925-29 to around 62 pounds in 1971 (Fig. 10). The per capita consumption of white potatoes far exceeds that of sweet potatoes, but sweet potatoes have declined in use by a larger proportional amount—from 21 pounds in the late 1920's to less than 4 pounds in 1971. This decline was nutritionally consequential in terms of the contribution of sweet potatoes to vitamin A in the diet.

Use of fresh white potatoes likewise declined—from around 147 pounds per capita in 1925-29 to about 58 pounds in 1971. However, the use of processed white potatoes has increased vastly since the post-War years of 1947-49 and now totals over 60 pounds per person. As a result, the consumption of white potatoes, both fresh and processed, totaled almost 120 pounds in 1971, up slightly from the late 1940's and only 25 pounds less than in 1925-29.

Processed sweet potato consumption has increased only slightly in the past four decades. Consequently, with the sharp decline in fresh use during this period, the per capita use of all sweet potatoes declined from over 21 pounds in 1925-29 to only 5 pounds in 1971.

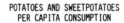

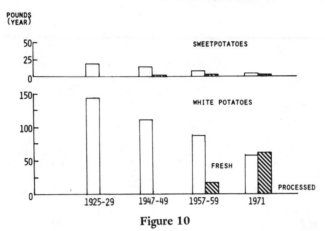

Figure 10

The trend in per capita consumption of fruits shows a pattern similar to that for vegetables, namely, a decline in the use of the fresh form and an increase in processed forms, including juices (Fig. 11). The average annual per capita consumption of fresh fruit declined from about 141 pounds in 1925-29 to 80 pounds in 1971, but the use of processed fruit increased from 38 pounds to 117 pounds in the same period. Thus, the total usage of fruits per capita increased 18 pounds over this period, from 179 pounds to 197 pounds.

Within these overall figures, however, the increased use of citrus is of special significance. Per capita consumption of citrus is of special significance. Per capita consumption of citrus increased from 33 pounds to 97 pounds per person over this period of nearly 50 years. There was a very pronounced shift to the use of processed forms, principally juices, while the use of fresh citrus declined slightly over this period. The decline in the use of fresh citrus since the post World War II years of 1947-49 was especially pronounced, from 55 pounds per capita to less than 30 pounds.

It should be noted here that information on per capita consumption of a specific product has limited usefulness in the deduction of nutrient intake—i.e., one-third of the consumers may use two-thirds of the product.

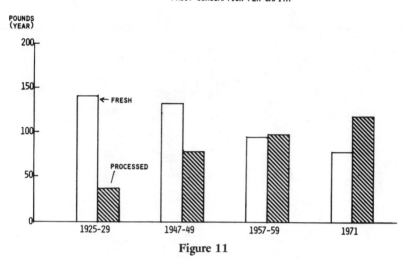

Figure 11

NUTRITIONAL TREND ACCOMPANIED BY INCREASED USE OF PROCESSED FRUITS AND VEGETABLES

As noted earlier, losses of some nutrients occur in connection with the processing and storage of processed fruits and vegetables. The nutritional trend that accompanied the greater use of processed items may be observed by comparing the trend in per capita consumption of fruits and vegetables with the trend in nutrients obtained from the fruit and vegetable group. Also affecting the trend in nutrients obtained from fruits and vegetables are shifts in selection among the fruit and vegetable group, both fresh and processed.

The average amount of vitamin A obtained from fresh fruits and vegetables per person, per day, declined over the last 50 years while the amount of this vitamin from processed forms increased more than three-fold (Table 3). In total, however, the amount of vitamin A obtained from fruits and vegetables declined from 4,330 I.U. in 1925-29 to about 3,860 I.U. in 1971 (Fig. 12). This decline is explained to a large extent by the substantial decline in consumption of sweet potatoes and dark green and yellow vegetables.

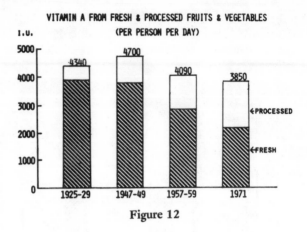

Figure 12

Table 3.—Nutrients Supplied by Fresh and Processed Fruits and Vegetables per Person per Day

Nutrient	1925–29	1947–49	1957–59	1971
Vitamin A (I.U.)				
Fresh--------------------	3,880	3,740	2,810	2,300
Processed----------------	450	960	1,280	1,560
Total--------------------	4,330	4,700	4,090	3,860
Thiamin (Mg.)				
Fresh--------------------	.35	.31	.25	.22
Processed----------------	.03	.06	.09	.13
Total--------------------	.38	.37	.34	.35
Vitamin B_6 (Mg.)				
Fresh--------------------	.77	.65	.55	.50
Processed----------------	.05	.09	.12	.16
Total--------------------	.82	.74	.67	.66
Ascorbic acid (Mg.)				
Fresh--------------------	96	91	71	63
Processed----------------	5	16	28	42
Total--------------------	101	107	99	105
Iron (Mg.)				
Fresh--------------------	2.9	2.5	2.0	1.8
Processed----------------	.6	1.0	1.3	1.6
Total--------------------	3.5	3.5	3.3	3.4
Magnesium (Mg.)				
Fresh--------------------	95	84	68	61
Processed----------------	10	17	22	26
Total--------------------	105	101	90	87

The significant trend to processed fruits and vegetables, principally citrus fruits and juices, has not resulted in a decline in the amount of vitamin C obtained from fruits and vegetables as a whole (Fig. 13). This is explained in part by the high level of retention of ascrobic acid in processed citrus products and by the commercial fortification of some of the fruit drink products in recent years.

The contribution of iron from fresh fruits and vegetables declined more than one-third over the span of 1925-29 to 1971 (Fig. 14). However, this was nearly compensated for by the increased amount contributed by processed forms. Consequently, the amount of iron obtained from all fruits and vegetables per person remained stable throughout the time period.

For thiamin, vitamin B_6 and magnesium, the trends to processed vegetables and away from vegetables that are good sources of these nutrients affected the contribution that fruits and vegetables made to the daily intake of these nutrients (Figs. 15, 16, 17). Although per capita consumption of all forms of fruits and vegetables increased, the amount of thiamin they provided declined by almost 10 percent and their contribution of vitamin B_6 and magnesium by almost 20%.

TRENDS OBSERVED IN USDA SURVEYS

Since 1936, the U.S. Department of Agriculture has made periodic nationwide food consumption surveys—in 1936, 1942, 1948 (urban only), 1955, and 1965-66. It is not possible to compare diet quality over the whole span of these surveys since changes were made in survey methodology. Also the Recommended Dietary Allowances (RDA), which served as standards against which to judge dietary levels, were revised several times in the past 30 years. Nevertheless, it was possible, using 1963 RDA's, to show comparisons between the spring 1955 survey and the spring 1965 portion of the most recent survey, which actually spanned a 12-month period, from April 1965 through March 1966.*

* For additional information, see Household Food Consumption Survey 1965-66, report number 6, "Dietary Levels of Households in the United States, Spring 1965," U.S.D.A., Agricultural Research Service.

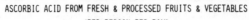

ASCORBIC ACID FROM FRESH & PROCESSED FRUITS & VEGETABLES
(PER PERSON PER DAY)

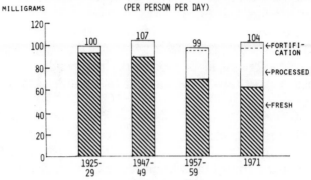

Figure 13

IRON FROM FRESH & PROCESSED FRUITS & VEGETABLES
(PER PERSON PER DAY)

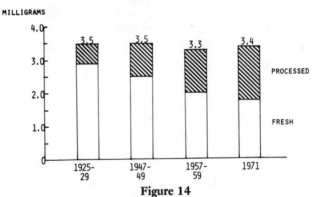

Figure 14

THIAMIN FROM FRESH & PROCESSED FRUITS & VEGETABLES
(PER PERSON PER DAY)

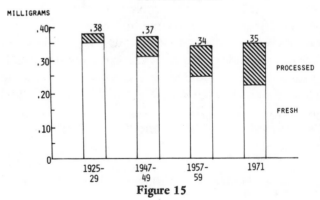

Figure 15

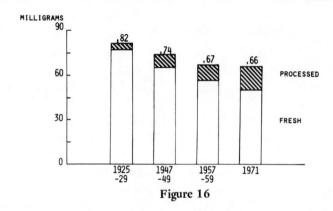

Figure 16

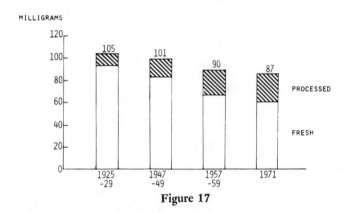

Figure 17

Fewer households had diets that met the RDA in 1965 than ten years earlier, according to these surveys. The nutrients most often found below the RDA in 1965 were vitamin A, ascorbic acid and calcium. It is significant to note that ascorbic acid and vitamin A, two of the nutrients found to be a problem in the 1965 survey, are nutrients which are supplied in significant amounts by fruits and vegetables.

Clues to the cause of lower levels of these key nutrients are shown by the profile of foods that were used in smaller a-mounts and those that were used in larger amounts per person in 1965 than in 1955. The consumption of vegetables and fruits, in terms of weight as purchased at the retail level, was 9% lower by households surveyed in 1965 than by those surveyed in 1955 (Fig. 18). The use of less milk and milk products, the main dietary sources of calcium, was undoubtedly associated with the lower levels of this nutrient in 1965. It appears that some of the milk and the fruits and vegetables in diets in 1955 were replaced by soft drinks and punches, and meat, poultry and fish in 1965.

Examination of the differences in the use of various forms of fruits and vegetables provides further indication as to causes of shifts in nutrient levels (Fig. 19). Households in 1965 used 17% less fresh vegetables and 14% less fresh fruit per person than households in 1955 (Table 4). Canned vegetable usage was 14% higher, and canned fruit usage 5% higher in 1965 than in 1955. Frozen vegetables and fruits, used in much smaller amounts than canned ones, increased in use by a greater proportion—

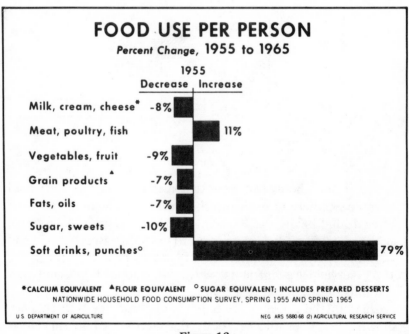

Figure 18

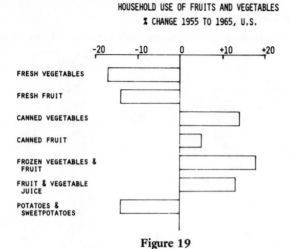

Figure 19

about 18% between 1955 and 1965. Nineteen percent less fresh potatoes, but triple the amount of frozen potatoes, were used in 1965 than were used 10 years earlier. On a fresh equivalent weight basis, there was almost no difference between the amounts of potatoes used in the two periods.

Closer examination of the use of good sources of ascorbic acid and vitamin A shows some important differences (Fig. 20). The 35% smaller quantity of fresh citrus fruits used by households in 1965 than ten years earlier could have severely jeopardized diet quality if the change had not been almost completely offset by the increased use of fresh and frozen citrus juices with their excellent retention of ascorbic acid. The lower consumption of fresh potatoes in 1965, mentioned earlier, and of dark green leaves and cabbages—all good sources of ascorbic acid— suggests a partial explanation for the lower quality of diets with regard to this nutrient. The smaller amounts of dark green leaves, carrots and sweet potatoes used in 1965 undoubtedly contributed to the lower quality of diets with regard to vitamin A. The slightly smaller amount of fresh tomatoes used was partially offset by increased use of canned tomatoes between 1955 and 1965.

If we hope to improve the dietary intake of problem nutrients by encouraging greater use of fruits and vegetables, we

Table 4.—Change in Household Use of Fruits and Vegetables, 1955-65 Per Household Per Week, U.S.

Items	Pounds per week		Percent change
	1955	1965	
Potatoes, sweetpotatoes	6.23	5.37	-14
Fresh white	5.82	4.71	-19
Fresh sweetpotatoes	.19	.11	-42
Commercially frozen	.04	.17	+325
Chips, sticks	.12	.22	+83
Fresh vegetables	8.86	7.33	-17
Dark green leafy	.52	.35	-33
Carrots	.65	.52	-20
Tomatoes	1.18	1.09	-8
Cabbage	1.00	.65	-35
Lettuce	1.20	1.30	+8
Corn	.54	.51	-6
Onions	.97	.82	-15
Fresh fruit	9.52	8.20	-14
Citrus	3.89	2.51	-35
Apples	1.30	1.38	+6
Bananas	1.45	1.45	---
Commercially canned vegetables, fruit	4.09	4.52	+11
Vegetables	2.58	2.93	+14
Fruit	1.51	1.59	+5
Commercially frozen vegetables, fruit	.56	.66	+18
Vegetables	.46	.62	+35
Fruit	.10	.05	-50
Juice: Vegetable, fruit (single-strength equivalent)	3.50	3.97	+13
Canned vegetable	.60	.54	-10
Canned fruit	1.51	1.41	-7
Frozen fruit	.36	.42	+17
Fresh fruit	.12	.46	+283
Dried vegetables, fruit	.61	.47	-23
Vegetables	.43	.37	-14
Fruit	.18	.10	-44

must first identify the households in which increased consumption would help to upgrade the diets. As one might expect, households with high incomes tended to use more fruits and vegetables (Table 5). The greater use of fresh forms was particularly noticeable among households with incomes over $10,000 (Fig. 21). The tendency of very low-income households to also use more fresh vegetables and fruit may be because these households were more likely to be farm households than households with higher incomes. The use of processed fruits and vegetables increased steadily from the lowest to the highest income level.

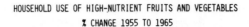

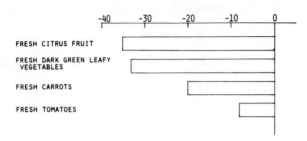

Figure 20

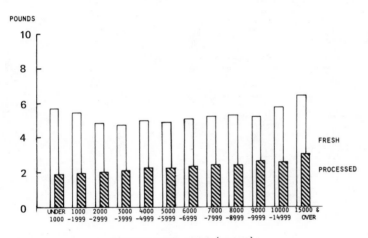

Figure 21

Categorizing households by urbanization revealed considerable differences in fruit and vegetable usage. As expected, rural
farm households used the greatest amount, averaging about 8.4
pounds per person in a week as compared to 7.5 for urban
families and 7.6 for rural nonfarm families (Fig. 22). Farm
families used significantly greater quantities of fresh fruits and
vegetables, including produce which they used for home can-

Table 5. Use of Fruits and Vegetables[1] per person in a week by Household Income, Season, Region, and Urbanization

U.S., 1965-66 12-month survey

Category	Total	Fresh [2]	Processed
	Lbs.	Lbs.	Lbs.
Household income (Year)			
Under $1,000--------------------------	7.52	5.61	1.91
$1,000-$1,999--------------------------	7.41	5.45	1.96
$2,000-$2,999--------------------------	6.89	4.81	2.08
$3,000-$3,999--------------------------	6.94	4.76	2.18
$4,000-$4,999--------------------------	7.34	5.02	2.32
$5,000-$5,999--------------------------	7.28	4.95	2.33
$6,000-$6,999--------------------------	7.50	5.10	2.40
$7,000-$7,999--------------------------	7.80	5.27	2.53
$8,000-$8,999--------------------------	7.85	5.34	2.51
$9,000-$9,999--------------------------	7.91	5.22	2.69
$10,000-$14,999-----------------------	8.45	5.82	2.63
$15,000 and over----------------------	9.65	6.54	3.11
Region (Year)			
Northeast-----------------------------	7.69	5.12	2.57
North Central-------------------------	7.46	5.12	2.34
South---------------------------------	7.59	5.37	2.22
West----------------------------------	7.95	5.38	2.57
Urbanization (Year)			
Urban---------------------------------	7.54	5.03	2.51
Rural nonfarm-------------------------	7.64	5.39	2.25
Rural farm----------------------------	8.36	6.67	1.69
Season			
Spring--------------------------------	7.16	4.72	2.44
Summer--------------------------------	9.30	7.27	2.03
Fall----------------------------------	7.06	4.58	2.48
Winter--------------------------------	6.94	4.30	2.64
All households			
Total U.S.----------------------------	7.64	5.24	2.39

1/ Excluding potatoes and sweetpotatoes.
2/ Includes home canned and frozen.

ning and freezing, than households in urban or rural nonfarm places. Urban households, on the other hand, used more commercially processed fruits and vegetables than the rural farm or rural nonfarm households.

In the 1965 survey, data were collected over a 12-month period in order that seasonal variations might be observed. As one would expect, the use of fruits and vegetables averaged over two pounds per person higher in summer than in other seasons (Fig. 23). The increase in summer consumption is accounted for entirely by an increased use of *fresh* fruits and vegetables. It is

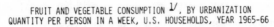

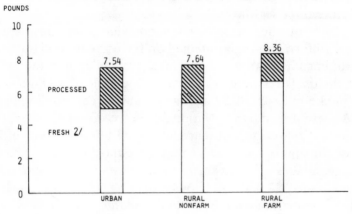

Figure 22

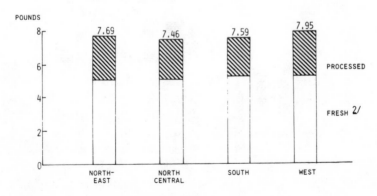

Figure 23

surprising to note, however, that about the same average quantity, including both the fresh and processed, was used in each of the other three seasons.

As part of the spring 1965 nationwide food consumption survey, information was obtained on the one-day food intake of approximately 15,000 men, women, and children in the survey households. Diets were evaluated on the basis of average nutrient intake for each of 22 sex-age groups, as compared with the RDA's established by the National Research Council in 1968. In general, the average nutrient intakes of females were further below the appropriate RDA's than were those of males, and the average intakes of individuals in households with annual incomes under $3,000 were lower than those of individuals in higher income households. Vitamin A and ascorbic acid were among the nutrients most frequently found to be below RDA levels of intake.

Survey results indicated relatively low intake levels of dark green and deep yellow vegetables. Even those sex-age groups with the highest usage averaged less intake than 25 grams of these high vitamin A vegetables in a day (Fig. 24). The USDA's

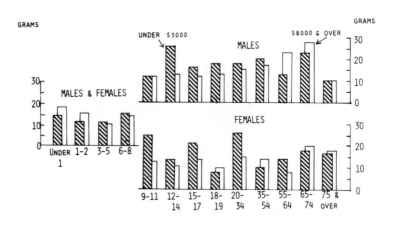

DARK GREEN AND DEEP YELLOW VEGETABLES
QUANTITY PER PERSON IN A DAY, BY INCOME

AGE IN YEARS

DIETS OF MEN, WOMEN AND CHILDREN, 1 DAY IN SPRING, 1965

Figure 24

Daily Food Guide recommends at least a serving—about 100 grams—of these foods every other day. Except for infants and older persons, individuals in low-income households consumed more dark green and deep yellow vegetables than persons in the higher income groups.

The intake of tomatoes and citrus fruits averages somewhat higher than the use of dark green and deep yellow vegetables. Income was found to be related to use of tomatoes and citrus. In all sex-age groups, persons in the higher income households, $8,000 and over, consumed more than those with household incomes of less than $3,000 (Fig. 25).

In summary, per capita consumption of all fruits and vegetables is greater today than it was five decades ago but is down slightly from the peak levels reached in the mid-1940's. Although total consumption has increased, there have been substantial shifts in consumer preference for certain fresh fruits and vegetables and for those that are processed. The use of both fresh fruits and vegetables declined sharply. This decline, however, was more than offset by the increased use of processed items.

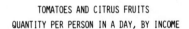

TOMATOES AND CITRUS FRUITS
QUANTITY PER PERSON IN A DAY, BY INCOME

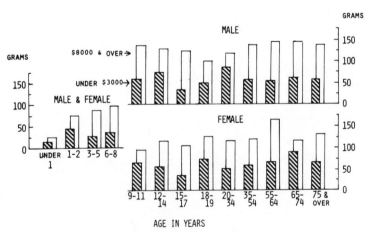

AGE IN YEARS

DIETS OF MEN, WOMEN, AND CHILDREN, 1 DAY IN SPRING, 1965

Figure 25

The shifts from the use of fresh fruits and vegetables to processed, as well as changes in selection among different fruits and vegetables, have resulted in some significant trends in the nutrients obtained from this food group. The amount of vitamin A obtained from fruits and vegetables has declined 11% since 1925-29 and 18% since 1947-49. Vitamin B_6 and magnesium declined by nearly 20% since 1925-29 while the amount of thiamin obtained from fruits and vegetables declined almost 10%.

It appears that increased consumption of fresh fruits and vegetables, particulary the high-nutrient forms, would be beneficial for many persons in need of dietary improvement. Educating consumers, particularly those with low incomes, to take greater advantage of the most economical and most nutritious fruits and vegetables, would offer a great potential for dietary improvement.

Acknowledgement

The authors wish to thank Louise Page and Berta Friend for their helpful comments.

Fruits and Vegetables: USDA Research for Tables of Food Composition*

Bernice K. Watt, Ph.D., Elizabeth W. Murphy, S.M., and Susan E. Gebhardt

Four aspects of our research in preparing food composition tables, with particular attention to fruits and vegetables, are discussed here. These aspects are (1) the background of the tables; (2) procedures and objectives in deriving the nutritive values; (3) analytical methods used and reliability of the data; and (4) current and planned work.

HISTORICAL BACKGROUND

USDA tables with data on composition of a wide range of foods were first issued in 1892. That year preliminary data on the composition of foods needed for evaluating diets were tabulated in one of the parts of the Report of the Connecticut Agricultural Experiment Station. The part that included the data on food composition was called "Investigations Upon the Chemistry and Economy of Foods." It was written by W. O. Atwater and C. D. Woods.[1] In 1928 a table devoted entirely to fruits, Department Circular 50[2] was issued, and in 1931, one devoted to vegetables, Department Circular 146[3] was issued.

During the first 50 years, data for the food items in the tables of composition were in terms of proximate composition.** Tables issued in 1945[4] and since have included data for proximate

---

* Official information material, U.S. Government. This chapter may be reproduced in part or in full.

** Proximate composition includes water (moisture), fat, protein, carbohydrate, and total ash.

composition and values for mineral elements and vitamins. The present *Agriculture Handbook 8*[5] includes data for 6 mineral elements—calcium, phosphorus, iron, sodium, potassium, and to a more limited extent, magnesium. The *Handbook* also includes data for 5 vitamins—vitamin A, thiamin, riboflavin, niacin, and ascorbic acid. Data for 3 additional B-vitamins—pantothenic acid, vitamin B_6, and vitamin B_{12}—were reported in a more recent publication, HERR 36, issued in 1969.[6]

From a fairly simple beginning in which carefully determined results obtained primarily from State Agricultural Experiment Stations were tabulated, problems in development of the tables of food composition have increased in number and complexity. Especially since World War II, the number of nutrients for which data must be reported has dramatically increased. Also, there has been a tremendous increase in the number of food items for which data must be tabulated. In recent years, we have also seen rapid changes in technology at each step in the food chain, from production through processing, storage, and preparation for serving. These changes have resulted in a need to review on a continuing basis the available data on food composition, in order to identify significant changes in nutrient content accompanying these developments.

OBJECTIVES AND PROCEDURES IN DERIVING
VALUES FOR PUBLICATION

From the first, two of the primary uses made of the data in the USDA tables have been for the nutritional evaluation of data obtained from studies of food consumption and for the nutritional assessment of this country's food supplies. Over the years the tables have come to be needed for additional and more specific purposes—for planning or evaluating normal diets and diets for special conditions, for developing food programs to improve nutritional status, for comparing nutritive values of different types of food and different forms of one kind of food, and for undertaking much research in food and nutrition.

In developing the data for the tables our objectives are to derive values that are representative for a food as used throughout the country on a year-round basis, to develop a data base that will permit us to revise the published data as circumstances require, and to provide information and additional data as needed to meet the increasing number of important special uses made of the tables.

For raw fruits and vegetables, as with other food groups, the first step in the process of developing data for our present tables is to locate suitable information. Search is made of scientific and technical journals and of special reports, and for unpublished work that investigators are willing to supply. This process of gleaning data is a continuous one. As data are assembled from any source they are screened, that is, given preliminary evaluation for accuracy and reasonableness. The data retained are entered on our special record cards. The analytical methods used in the analyses and several kinds of information about the sample are also recorded if available. Often it is necessary to communicate directly with the author to obtain supplemental details, as published reports do not always specify part or parts of the product analyzed, degree of maturity, procedures used in preparing the sample for analysis, or other details of concern in evaluating data for use in food composition tables.

The next step, a very critical one, comes at the time a new revision or a new publication is planned. Assembled information and all recorded data—those used in a previous edition if there has been one and those recorded more recently—are reviewed for reliability and for current significance.

The factors and considerations that enter into the procedures in deriving representative values are not the same for each food or for each group of nutrients. Fresh produce may be markedly affected by developments in production and marketing practices. Some of these developments significantly alter the levels of one or more major nutrients present. In deriving values for fresh fruits and vegetables, attention has been directed particularly to those nutrients which are labile or otherwise subject to change and which are provided in important amounts by these foods. In addition to the differences in the nature of the various fruits and vegetables and the variation in stability of nutrients, there is the important problem of the kinds and extent of data available for use in deriving updated values.

Reviews of available data characteristically show a wide range in the array of values reported for several of the more important nutrients in most fruits and vegetables. Procedures used by investigators in preparing and analyzing samples are reviewed for any possible effect that differences in methodology might have on analytical results. For a variety of reasons some of the data that have been entered on the file cards may be eliminated. If enough data remain, summaries are made of the values classified by any factors that would appear to be related to content of the nutrient being studied. Included would be such factors as variety, location, year or period of harvest, degree of maturity, and condition of storage.

Along with data on nutrients, information on recent or current trends in production and marketing practices is sought from government publications, specialists within the Department of Agriculture, and industry groups. This information is used in conjunction with the data on composition in deriving average values for nutrients appropriate for the food as currently marketed.

Ascorbic acid values for citrus fruit and potatoes will serve to illustrate some of the procedures used in deriving data in *Handbook 8*. As preliminary review indicated there might be some varietal differences, data selected for use in the *Handbook* were limited to the important commercial varieties. For oranges, the data used were further limited to samples meeting state standards for maturity for marketing.

As shown in Figures 1 and 2, different kinds of citrus fruit are not equally rich sources of vitamin C, and some varieties undergo considerable change in composition with a marked drop in ascorbic acid as their harvesting season progresses.

Oranges, like other citrus fruits, are a dependable source of ascorbic acid. They retain this value well during storage. Nevertheless, a wide range in content is found among individual samples. Results from thousands of analyses have shown that values may range from less than 10 to more than 80 mg of ascorbic acid in 100 gm of juice. Variety and period in the harvest season account for some of this difference. Figure 2 shows the average values found for ascorbic acid in juice from oranges of the important commercial varieties at different periods in the harvest season.

SEASONAL CHANGES IN ASCORBIC ACID OF SOME CITRUS FRUITS

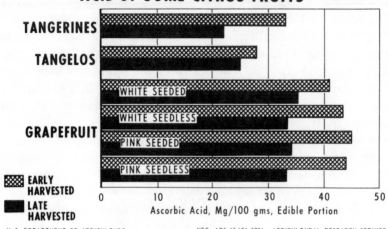

U. S. DEPARTMENT OF AGRICULTURE   NEG. ARS 63 (9) 5736   AGRICULTURAL RESEARCH SERVICE

Figure 1

SEASONAL ASCORBIC ACID CONTENT OF ORANGES

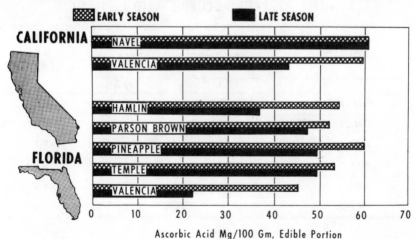

U. S. DEPARTMENT OF AGRICULTURE   NEG. ARS 63 (9) 5734   AGRICULTURAL RESEARCH SERVICE

Figure 2

Using information of the type shown in Figure 2, and statistics on shipments of oranges for use as fresh fruit, a weighted average value was developed for each of the commercially important varieties of oranges reaching the market. The weighted average value for ascorbic acid in Florida Valencias was found to be 37 mg of ascorbic acid per 100 gm of juice and is shown in Figure 3. The statistics on shipments used were based on the period 1953-56. A check made in 1961 as the manuscript for the *Handbook* went to press indicated that no change in the figures on shipment was needed at that time. Florida Valencias are a late-season variety. The weighted average value for Florida oranges including Valencias and other important commercial varieties was found to be 45 mg of ascorbic acid per 100 gm of juice.

Data for ascorbic acid in the predominating varieties of oranges grown in California, Valencias and Navels, were also summarized by month of harvest. Again, by applying statistics for monthly shipments of oranges for use as fresh fruit, average values were calculated and found to be 49 and 61 mg per 100

FLORIDA VALENCIA ORANGES
THEIR AVERAGE ASCORBIC ACID CONTENT

MONTH	TOTAL SHIPMENTS*	ASCORBIC ACID	WEIGHTED BY SHIPMENTS
	Pct.	Mg./100 gm.	Mg.
FEBRUARY	9	44.7	4.0
MARCH	26	39.2	10.2
APRIL	24	37.9	9.1
MAY	24	37.2	8.9
JUNE	13	31.0	4.0
JULY-AUGUST	4	22.3	.9
TOTAL	100		37.1

*MONTHLY AVERAGE FOR 1953-56

U. S. DEPARTMENT OF AGRICULTURE   NEG. ARS 63 (9) 5733   AGRICULTURAL RESEARCH SERVICE

Figure 3

gm respectively for the 2 varieties. Using these values, and taking into account the relative proportions of the total orange supply coming from California and Florida, a countrywide, year-round value of 50 mg ascorbic acid per 100 gm juice was derived.

Further discussion of the problems in deriving representative data for nutrients in citrus fruit may be found in an earlier publication.[7]

Potatoes present problems that differ somewhat from those for oranges. Although present use of fresh potatoes has declined by about a third from a per capita consumption of nearly 90 lb. per year in the mid-1950's,[8] potatoes from the produce counter are still an important source of ascorbic acid in U. S. diets and contribute significant amounts of a number of other nutrients. Fewer data on ascorbic acid in well-described samples were found for potatoes than for oranges. The data available indicated that length of storage period had an important influence on content of ascorbic acid and that variety and degree of maturity might also affect the value.[9] When the data were separated by various categories, the ascorbic acid content was found to be highest in immature potatoes—those which are sometimes grabbled for home use before the crop is ready to dig. For the immature potatoes the average value found was 35 mg per 100 gm. As these potatoes do not get into commercial channels, data for them were not used in deriving the weighted average value shown for potatoes in the *Handbook.*

For potato varieties of commercial importance, the average values within 2 weeks after digging were found to range from 19 to 33 mg of ascorbic acid per 100 gm, with 26 mg a reasonable figure for new potatoes on the regular market.

Figure 4 shows that potatoes lose ascorbic acid rather rapidly during early months of storage. Since the most rapid losses were observed for the varieties with highest initial concentrations, the differences among varieties decreased during storage.

For each of 8 varieties, ascorbic acid contents of potatoes stored for different periods of time were calculated as percent of content for freshly harvested tubers. Results are shown in Figure 5. After 1 month of storage, the ascorbic acid had dropped to about three-fourths the original level; after 3

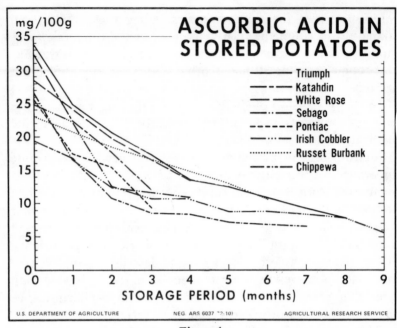

Figure 4

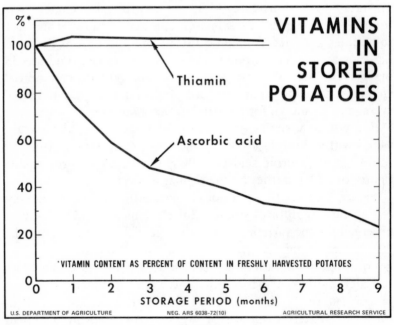

Figure 5

months to only half; and after 6 months to about a third. After 9 months of storage the ascorbic acid content of potatoes was only about a fourth as high as in freshly harvested potatoes.

Thiamin, unlike ascorbic acid, is apparently well-retained by potatoes during storage. [10] No adjustment for storage loss was made in the thiamin values in potatoes.

For each month of the year an average value for ascorbic acid in potatoes was calculated. The basis for the average ascorbic acid value using data for the month of December is illustrated in Table 1. The average value for each month was weighted to take into account variety and length of storage and the proportions of stored potatoes and new potatoes. The weighted average values found for each month throughout the year are shown in Figure 6. The distribution in the proportion of potatoes marketed as produce each month throughout the year is also shown in this figure. These data were used with the weighted average values for ascorbic acid by month to arrive at the composite value of 20 mg of ascorbic acid per 100 gm, the value we have

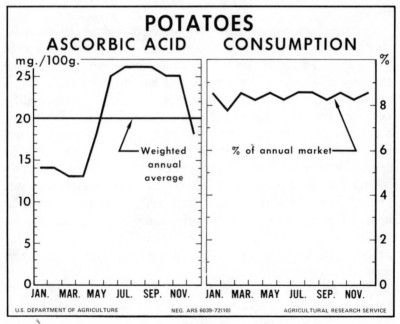

Figure 6

Table 1   ASCORBIC ACID IN POTATOES
                              MARKETED IN DECEMBER

VARIETY	TOTAL MARKET	ASCORBIC ACID	WEIGHTED BY TOTAL MARKET
STORED:	%	mg./100 g.	mg./100 g.
Katahdin	46.6	17.0	7.9
Reds	18.7	17.5	3.3
Russet	18.7	15.5	2.9
Other White*	9.3	16.5	1.5
NEW:			
White Rose	6.7	28.4	1.9
TOTAL	100		17.5 18

*Round
                              DATA FOR 1954
U.S. DEPARTMENT OF AGRICULTURE
NEG. ARS 6042-72(10)   AGRICULTURAL RESEARCH SERVICE

used for a year-round weighted average for potatoes marketed commercially.

For the 2 examples used here, oranges and potatoes, some changes have occurred in production and marketing practices in the 12 to 15 years since the data were developed for the *Handbook*. Except for Katahdin and Russet Burbank, varieties of potatoes that were important then have been supplanted by new varieties. Introduction of new varieties does not necessarily mean change in levels of nutrients, but some exploratory work is needed to determine whether or not the values continue to be applicable. Retentions of ascorbic acid during storage probably follow the same pattern for newly important varieties as was observed for the varieties just discussed. However, we need at least a few analyses on these new varieties, using samples for which storage time is known, to show whether or not they differ significantly in ascorbic acid content from varieties formerly important.

For sweetpotatoes, a major shift in the varieties produced has occurred. We expect the shift to have a very marked effect on the vitamin A value, possibly also on the vitamin C value.

Limited data indicate that Centennial, a relatively new variety and one that is in heavy production, has a much higher vitamin A value than had most of the kinds of sweetpotatoes previously grown for commercial production.

Mineral element values shown in the *Handbook* for fruits and vegetables, like vitamin values, have been based on data assembled from many sources. On the whole, fewer samples of each kind of fresh produce have been analyzed for content of mineral elements than for content of ascorbic acid or a few of the other important vitamins. Also, many of the data for mineral element contents are comparatively old. For very few fruits or vegetables were usable data for any one element available from 100 or more samples. Frequently, the number of samples analyzed has totaled fewer than 35 samples of one kind of fruit or vegetable; this includes data reported as long ago as the 1920's.

More uniformity among data reported for mineral element contents might be expected than for vitamin contents, as mineral elements are not vulnerable to losses as can occur in vitamin values when produce is bruised, exposed to light, subjected to elevated temperatures, or held for varying periods of time. Wide ranges were observed, however, in data for mineral elements reported for different samples of each kind of food. Possibly even wider ranges might have been observed if sampling had been more extensive.

Much remains to be learned about factors relating to different levels of mineral elements in fruits and vegetables. Geographic area of production and the associated environmental conditions may have a significant effect. Some studies have been made of the effect of soils, fertilizers, and other climatic factors under controlled conditions on content of mineral elements in selected food crops such as cabbage, beans, carrots, broccoli, and turnip greens. [11-13] These studies have generally shown variability to be as great or greater within a location as between locations. Wide variability for certain mineral elements was found both within and between sites. Sodium was particularly variable. Iron, magnesium, and potassium were variable, but less so than sodium. Calcium and phosphorus showed comparatively little variability.

Research in Florida [11] on fresh vegetables and in Hawaii [13] on fresh fruits and vegetables has shown that sodium content of the crops was markedly influenced by sodium in the soil. Sodium can apparently be deposited in soils by irrigation water as well as by fertilizer, and tends to be higher in soils near the sea than in those inland.

In addition to sodium, at least two trace elements, selenium and iodine, are known to vary widely with geographic location. Selenium and iodine are two important trace elements for which tables of representative values must be prepared in the future. Guidance in evaluating geographic effects on their contents in foods is therefore being sought.

With the new, greatly improved methods of analysis now available for determining each of the mineral elements in foods, it is possible that new research may uncover other examples of variation with geographic location and better identify those factors that are related to variations in the levels of the mineral elements in foods.

METHODS OF ANALYSIS AND RELIABILITY OF THE DATA

An important part of the work of evaluating data for use in our handbooks is the review of analytical methods for accuracy, reliability, and suitability for use with the foods analyzed. Data for vitamins in the tables issued in 1945[4] were based to considerable extent on biological assays. Except for vitamin C, the measurements were frequently in terms of rat growth. Those time-consuming animal assays were abandoned as chemical, physical, and microbiological assays were introduced and developed into expedient, reproducible methods which permitted greater precision than did the early animal assays. During the transition period, however, the results from the early studies served a very important purpose. They provided a reference, almost a standard, for accepting the data being obtained by the newer methods. Any wide discrepancies observed served as a signal to investigate possible causes. Problems in sampling, for example, rather than method of analysis could have accounted for some discrepancies.

In the present edition of the *Handbook,* values for thiamin in fruits and vegetables are based mostly on analysis by the thiochrome procedure, although in some studies microbiological assays—particularly the yeast fermentation method—were used. For riboflavin, fluorometric and microbiological methods were usually used. Niacin values in the *Handbook* are for total preformed niacin and do not include any niacin value contributed by the amino acid, tryptophan, present in the food. Most investigators who reported data for niacin used microbiological procedures but some used chemical methods—usually the reaction with cyanogen bromide.

Ascorbic acid values for fresh fruits and vegetables in the *Handbook* have been based almost entirely on data for reduced ascorbic acid.

Vitamin A values listed in the present edition of the *Handbook* were derived mainly from physical-chemical determinations of total carotenoids or of individual carotenes and cryptoxanthin, whereas sources of data in the previous edition had included some data obtained by various biological methods.

We intended to separate the values for vitamin A in the present edition of the *Handbook* under different columns showing data separately for beta-carotene and other precursors of vitamin A. However, because of the limitations of the data available at the time this edition of the *Handbook* was prepared, it was not possible to show meaningful data for the different precursors of vitamin A. Even today, data are lacking for these precursors in many foods.

We believe the vitamin values shown for fruits and vegetables in the *Handbook* to be reasonably reliable and appropriate for the uses intended. That is, we believe them to represent the products available on a year-round basis to the consumer. There have been many problems in arriving at the published data, as each value has been derived from diverse sources and the individual investigators have used a variety of adaptations in methods for sample preparation and analysis. Often we have turned for counsel to experts on matters of methodology, especially to the research scientists in the Agricultural Research Service. Also, we have gone outside the Department to other government agencies, to academic sources and at times to research scientists in industry, for expert help with specific problems. We are most grateful for the fine cooperation we have had throughout the years from so many who have responded to our needs. This type of assistance in appraising the data has been invaluable and indispensable to us, just as has the information on production, handling, and marketing practices of food stocks provided by other experts within the Department and by seed and grower organizations.

In terms of methodology, the picture is probably brighter for analysis of mineral elements than for most other nutrients. In the past few years, atomic absorption spectroscopy has emerged into widespread use. This method is relatively inexpensive, rapid, and when properly handled, accurate. A considerable body of data on mineral element content of foods, based on use of this method, is beginning to accumulate. Other methods currently in use include flame photometry, emission spectrometry, and traditional chemical methods. A few data based on neutron activation are beginning to appear.

In examining the array of data available for some of the mineral elements in foods, a wide range of values is frequently seen. In a few instances, data are available as determined by two or more methods on the same sample. We have compared some of these data [14-26] and charted the results expressed as ratios. While the samples analyzed include very few fresh vegetables and fruits as such, it is likely that similar results would be obtained with produce.

Figure 7 shows the ratio of results obtained on the same sample by atomic absorption or chemical methods in relation to data obtained by emission spectrometry. Had the results been identical for a given comparison, the ratio would have been one. As the chart shows, both atomic absorption and chemical methods gave higher values than emission except for four comparisons involving zinc, manganese and potassium. For copper, the differences between methods were so great as to make us wonder if copper values obtained by emission have any usefulness. Emission values for iron and magnesium also would appear to be of concern. Of the foods shown in this chart, mixed diets and spices were analyzed in the same laboratory. Ten to 30 samples of the mixed diets and 6 samples of spices were in-

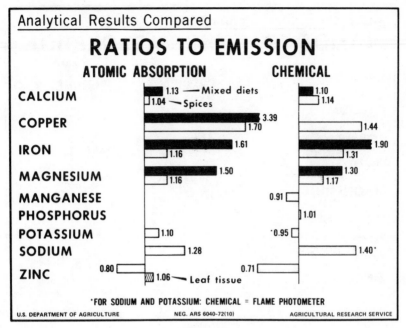

Figure 7

volved in the comparisons. Between the time analyses were made on the mixed diets and on the spices, the emission procedures were rigorously tested and modified. These modifications improved the method so that results were more in line with results obtained by other methods, but emission spectrometry still appears to be far less satisfactory than atomic absorption, flame photometry, or chemical methods for determining most elements.

Comparisons of results by chemical and flame photometric methods with atomic absorption are shown in Figure 8. In most instances, ratios show that the values obtained by each pair of methods agree within about ± 10%, indicating that the methods give reasonably interchangeable results. Such findings are especially reassuring when they are made in interlaboratory collaborative assays, as was the case for data shown in this chart for feeds. (Data for feeds are averages for 7 samples analyzed in 11 laboratories.)

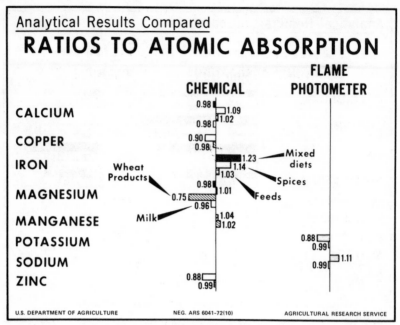

Figure 8

Some comparisons on this chart show rather poor agreement. The first is iron in mixed diets, comparing chemical and atomic absorption values. In this case, the analyst indicated that atomic absorption data were probably more accurate, as there was some question about the chemical standards. Data for iron in spices, although in closer agreement than data for the mixed diets, still show poorer agreement than we would like to see.

Other analyses in poor agreement were those of magnesium in wheat products. In this instance, the same 156 samples were analyzed, by chemical methods in one laboratory and by atomic absorption in another location. Here again we would judge the atomic absorption values to be more accurate. The AOAC chemical method used involves precipitation of calcium, as oxalate, with subsequent determination of magnesium on the filtrate. Possibly some magnesium precipitated along with the calcium.

Table 2 shows other comparisons of methods for a few mineral elements. Except for hard-to-sample materials such as

Table 2 **Analytical Results Compared**
PHYSICAL AND CHEMICAL METHODS

METHODS COMPARED	MINERAL ELEMENT	FOOD ANALYZED	RATIO OF RESULTS
Chemical/flame photometry	Ca	milk	1.01
	Na	milk	1.00
	Na	vegetables	.99
Flame photometry/K_{40}	K	lean meat	1.00
	K	fat	1.17
Atomic absorption/K_{40}	K	lean meat	.98
	K	bone	1.01
	K	fat	1.14

U.S DEPARTMENT OF AGRICULTURE
NEG. ARS 6043-72(10) AGRICULTURAL RESEARCH SERVICE

fat, the methods give essentially identical results. These data are especially reassuring. For many foods, sodium values in *Handbook 8* are based primarily on flame photometry analyses. Many potassium values are based on flame photometry and ^{40}K analyses. We have in our files a large grist of unpublished data on potassium in fruits and vegetables, which had been analyzed for ^{40}K as part of radioactivity monitoring programs. These data were made available to us as a spin-off of the monitoring program, and were of great importance in deriving values for inclusion in *Handbook 8.*

There is an urgent need for simple, reliable methods of analysis collaboratively tested and adapted as needed for use with a wide variety of foods.

PLANS FOR CONTINUING AND EXPANDING THE DATA BASE AND UPDATING THE TABLES

Currently we are reappraising data for vitamin A, food by food, for reliability and delineation of the problems in shifting from the terms that have been used—International Units of Vitamin A Value—to units of weight for the different vitamin

A-active components present. On the basis of progress made so far in this reappraisal, we believe the data for vitamin A for some fruits and vegetables, especially the red-, orange-, and yellow-colored products, to be less satisfactory than data for the green vegetables. Data obtained from analyses involving hot saponification will need to be reconsidered and possibly dropped from the array. Likewise, data for total carotenes reported as beta-carotene will need to be reassessed if subsequent analysis indicates that alpha-carotene or other precursors having lower activity than beta-carotene form a major part of the total carotenes present. We already know that the data for carrots need reassessing. We anticipate a change in values for some of the other vegetables and fruits, with relatively less change for the vitamin A values of green vegetables than for others.

In the discussion thus far we have not mentioned protein and amino acids, as fruits and vegetables have so little compared with most other foods. For certain therapeutic diets, the low levels of amino acids, in combination with significant contributions of minerals and vitamins in fruits and vegetables, are a distinct advantage. Emphasis is being given to including data for fruits and vegetables in the revision we have in preparation of our HERR No. 4, "Amino Acid Content of Foods."[27] However, few reliable data can be found for amino acids in these foods, in great part because satisfactory methods of hydrolyzing and analyzing such low-protein foods have not been developed and tested.

Our staff has published data for the content of vitamin B_6 in foods,[6] but much work remains to be done on this nutrient in fresh fruits and vegetables. We know practically nothing about losses of total B_6, or shifts among the forms of B_6, with storage of foods such as potatoes. Neither do we have information on varietal or seasonal differences in content of this vitamin, or of losses with cooking. The assay procedure for vitamin B_6 has been collaboratively tested, but is tedious and difficult. Reliable data are not always obtained for B_6 even from careful analysts. Until a simpler, more reliable method for determination of the separate forms of B_6 is available, it is unlikely that there will be much increase in knowledge about the forms and amounts of this vitamin in foods.

Work on a number of other projects is currently underway. Major emphases for our staff at present include revising our tables on food yields at different stages of preparation and compiling data on trace elements in foods. We have just completed a comprehensive table of nutrients in common household portions and market units of foods. In addition, our staff plans and monitors research done under contracts and grants. Most research we are sponsoring at present is on poultry and processed foods. We will also be obtaining badly needed data on mineral and proximate composition of dry beans and peas, and on nutrients in foods, including some fruits and vegetables, used in Hawaii. We are working toward a new revision of the *Handbook* and expect it to be in a new format—a loose-leaf format with one page devoted to a food, and data shown on the page for all nutrients. The advantages for updating and expanding the information with such a format are obvious.

We are continuing to search for data as we have in the past from published and unpublished sources, and to study the relationships among different forms of a food—raw, processed, and prepared. We will be having a new dimension in our work, however. We expect to have a very much greater input from the food industry than in the past, and to expand our present files into a Data Bank. The data we are already receiving from industry should help at several steps in our work but especially in supplying data for products that are mixtures of food items, including the convenience and specialty products that daily become more important items of consumption. There has never been a lull or a dull period in our work of developing tables of food composition. The task ahead now of updating and expanding the *Handbook* seems to be larger and much more challenging than ever before.

References

1. Atwater, WO and Woods, CD: Investigations upon the chemistry and economy of foods. Connecticut (Storrs) Agricultural Experiment Station 1891 Report, 1892.
2. Chatfield, C and McLaughlin, LI: Proximate composition of fresh fruits, Circular 50, U.S. Department of Agriculture, 1928.
3. Chatfield, C and Adams, G: Proximate composition of fresh vegetables, Circular 146, U.S. Department of Agriculture, 1931.

4. Tables of food composition in terms of eleven nutrients, Misc Publ 572, Bureau of Human Nutrition and Home Economics, U.S. Department of Agriculture, 1945.

5. Watt, BK and Merrill, AL: *Composition of foods . . . raw, processed, prepared; Agriculture Handbook 8,* U.S. Department of Agriculture, 1963.

6. Orr, ML: Pantothenic acid, vitamin B $_6$ and vitamin B $_{12}$ in foods, Home Economics Research Report 36, U.S. Department of Agriculture, 1969.

7. Merrill, AL: Citrus fruit values in "Handbook No. 8", revised, *J Amer Dietet Assoc* 44:264-270, 1964.

8. Seelig, RA: Fruit and Vegetable Facts and Pointers: Potatoes. United Fresh Fruit and Vegetable Association, 1972.

9. Leichsenring, JM; Morris, LM; Salmon, WD; et al: Factors influencing the nutritive value of potatoes, Technical Bulletin 196, University of Minnesota, 1951.

10. Yamaguchi, M; Perdue, JW; and MacGillivray, JH: Nutrient composition of White Rose potatoes during growth and after storage, *Amer Potato J* 37:73-76, 1960.

11. Janes, BE: Composition of Florida-grown vegetables. III. Effects of location, season, fertilizer level and soil moisture on the mineral composition of cabbage, beans, collards, broccoli, and carrots. Bulletin 488, Agriculture Experiment Station, University of Florida, 1951.

12. Speirs, M; Miller, J; Peterson, WJ; et al: Influence of environment on the chemical composition of plants, Bulletin 42, Southern Coop. Series, 1955.

13. Wenkam, NS; Miller, CD; and Kanehiro, Y: Sodium content of Hawaii-grown fruits and vegetables in relation to environment, *J Food Sci* 26:31-37, 1961.

14. Bennink, MR; Ward, GM: Johnson, JE; and Cramer, DA: Potassium content of carcass components and internal organs of cattle as determined by ^{40}K and atomic absorption spectrometry, *J Anim Sci* 27:600-603, 1968.

15. Christensen, RE; Beckman, RM; and Birdsall, JJ: Some mineral elements of commercial spices and herbs as determined by direct reading emission spectroscopy, *JAOAC* 51:1003-1010, 1968.

16. Clifford, PA and Winkler, WO: Report on the determination of sodium in foods, *JAOAC* 37:586-600, 1954.

17. Unpublished data from Nutrient Data Research Center, Consumer and Food Economics Institute, Agricultural Research Service, U.S. Department of Agriculture, Hyattsville, Maryland.

18. Heckman, M: Minerals in feeds by atomic absorption spectrophotometry, *JAOAC* 50:45-50, 1967.

19. Johnson, JE; Ward, GM; and Sasser, LB: Calibration techniques for a crystal-type whole-body counter, in *Body Composition in Animals and Man,* publ 1598, National Academy of Sciences, Washington, DC, 1968, pp 317-325.

20. Jones, JB: Elemental analyses of plant leaf tissue by several laboratories, *JAOAC* **52**:900-903, 1969.
21. Keirs, RJ and Speck, SJ: Determination of milk minerals by flame photometry, *J Dairy Sci* **33**:413-423, 1950.
22. Kirton, AH and Pearson, AM: Comparison of methods of measuring potassium in pork and lamb and prediction of their composition from sodium and potassium, *J Anim Sci* **22**:125-131, 1963.
23. Lohman, TG; Dieter, RC; and Norton, HW: Biological and technical sources of variability in bovine carcass lean tissue composition. I. Technical variation in measurement of potassium, nitrogen, and water, *J Anim Sci* **30**:15-20, 1970.
24. Martin, TG; Stant, EG, Jr; Kessler, WV; and Parker, HE: Association between chemical and counter estimates of potassium content, in *Body Composition in Animals and Man*, Publ 1598, National Academy of Sciences, Washington, DC, 1968, pp 341-349.
25. Murthy, GK and Rhea, U: Determination of major cations in milk by atomic absorption spectrophotometry, *J Dairy Sci* **50**:313-317, 1967.
26. Zook, EG; Greene, FE; and Morris, ER: Nutrient composition of selected wheats and wheat products. VI. Distribution of manganese, copper, nickel, zinc, magnesium, lead, tin, cadmium, chromium, and selenium as determined by atomic absorption spectroscopy and colorimetry, *Cereal Chem* **47**:720-731, 1970.
27. Orr, ML and Watt, BK: Amino acid content of foods, Home Economics Research Report 4, U.S. Department of Agriculture, 1957.

Nutritive Losses in the Home Storage and Preparation of Raw Fruits and Vegetables

Catharina Y. W. Ang, Ph.D. and G. E. Livingston, Ph.D.

Food composition tables are generally based on analytical data obtained on fresh, canned, frozen, dried or "cooked" food products as if each of these states were a clearly definable and unalterable constant. In so doing, the tables fail to deal with the more subtle differences involved, which may, in fact, in the case of some nutrients - particularly those which are affected by heat, oxidation or leaching - have a significant impact on the levels actually present. For example, data on fresh fruits and vegetables do not consider losses in home storage, washing or preparation. Data on canned fruits and vegetables do not distinguish between various size cans, or processing methods used although these factors - due to the different heat process requirements involved - are known to influence the retention of heat labile nutrients. Data on frozen fruits and vegetables do not specify the freezing method although thawing losses in conventionally frozen products might conceivably differ from those losses encountered with individually quick-frozen products. Demonstrable differences in nutrient losses are related to cooking methods involved, and those are not generally addressed by the usual food composition tables. Finally, losses which may occur when cooked foods are held hot for any length of time or are chilled and rewarmed are usually ignored too.

The purpose of this paper is to review the available data on nutrient losses which may occur in raw fruits and vegetables between the time when they are purchased by the homemaker in a retail store and the time they are consumed at the family table. Frozen as well as fresh products will be considered since, in spite of the blanching treatment received, frozen vegetables are essentially raw products as purchased.

In this perspective, losses of nutrients may occur at the following stages:

1. During storage in the home at ambient temperature, refrigerator temperature or freezer temperature.

2. During preparation of fruits and vegetables prior to cooking and/or serving, i.e., peeling, trimming, cutting, dicing, shredding, washing, water or brine immersion, sugaring or juice extraction.

3. During cooking, i.e., boiling (covered or uncovered), dry cooking, pressure cooking (steaming), frying, baking or microwave cooking; with the effects of adjuncts such as sugar, vinegar, salt, monosodium glutamate or baking soda.

4. During holding of cooked foods, while hot; or during chilling or the reheating of leftovers.

Data available in the literature leave considerable gaps in our knowledge of the effects of some of the factors cited. Furthermore, many of the early papers offer no statistical analyses to support the findings reported, and most papers dealing with the effect of cooking methods fail to report the time - temperature relationships involved. In the case of heat-labile nutrients, such as thiamin, temperature curves would permit a kinetic comparison to be made of the thermal destructiveness of the various methods used and would allow results from different investigators to be compared. Finally, there appears to be - probably as a result of its relative instability and ease of analysis - undue emphasis given to ascorbic acid, at the expense of other micronutrients.

NUTRIENT LOSSES INCURRED IN HOME STORAGE

Hewston et al[1] studied the effects of storage for up to 6½ months at 40° F. and 85% relative humidity on the ascorbic acid content of U.S. No. 1 Maine grown Katahdin potatoes. The ascorbic acid content was gradually reduced to about half of the original level but during the latter part of the storage period, when the potatoes had begun to sprout, there was an increase in ascorbic acid. Interestingly, potatoes withdrawn during the first week of storage and boiled whole, unpared retained an average of 75% ascorbic acid compared with 99% throughout the rest of the storage period.

The authors speculated that the increasing stability of ascorbic acid as related to storage time might be due to changes in the enzyme system or the development of suberization. Von Loesecke[2] summarized data published by a number of investigators from 1937 to 1947 which indicated storage losses of ascorbic acid of up to 75%. At least one of the early studies suggested that, in the two varieties studied (Russet Burbank and Bliss Triumph), ascorbic acid losses over a 6 month storage period were lower when potatoes were stored in a warm, dry cellar at 55° - 60° F. (0-20% loss) than in a cool, damp cellar at 37° - 40° F. (30-50% loss). Bring et al[3,4] measured ascorbic acid losses in stored Russet Burbank potatoes. In one study, in which potatoes were stored at temperatures of 38°-54° F., the ascorbic acid content of the raw potatoes on a dry-weight basis declined from 121.3 mg/100 gm to 52.1 mg/100 gm after about 4 months and to 45.3 mg/100 gm after 7 months. In another study the ascorbic acid content of potatoes stored 6 months at 39° to 40° F. was reduced from 123.2 mg/100 gm to 53.0 mg/100 gm. Page and Hanning[5] studied the effects of storage time and temperature on the niacin and pyridoxine content of several varieties of potatoes. A total of 147 samples of raw, peeled potatoes were analyzed for niacin and vitamin B_6 content. On a wet basis, niacin ranged from 1.03 to 2.08 mg/100 gm and vitamin B_6 from 0.13 to 0.42 mg/100 gm. Storage of Wisconsin Cobblers and Triumph potatoes at 40° F. for up to 6 months showed an initial increase in niacin for the first month, followed by a gradual decline so that at the end of 6 months the niacin content was approximately the same as at the time of harvest. The study indicated a possibility of greater destruction of niacin in potatoes held at room temperature (75° F.) that in those held cold (40° F.). Vitamin B_6 levels increased 152% in Cobblers and 86% in Triumphs in 6 months at 40° F., these increases being greater than what could be attributed to weight loss in storage. In some varieties, the vitamin B_6 content of potatoes stored at room temperature increased more rapidly than in those stored at 40° F.

Storage studies on carrots were also carried out by Hewston et al[1] using Chantenay variety carrots. The ascorbic acid content of carrots, unlike that of potatoes, was unaffected by storage for 2 months. A summary of other studies (reported between 1935 and 1949) on 7 other vegetables was prepared by von Loesecke.[2] This summary shows ascorbic acid losses ranging from 0% (unhusked corn held refrigerated for 7 days) to 95% (spinach held 192 hours at room temperature). The following ascorbic acid losses were reported for vegetables held for 24 hours under cool or refrigerated temperature conditions:

Asparagus	- 30°F.	- 3% loss
Green beans	- 46-50°F.	- 10%
Broccoli	- 46-50°F.	- 10-30%
Swiss chard	- 46-50°F.	- 30%
Spinach	- 41°F.	- 19%

Fruits are generally not stored in the home for more than a few days and few data are available to permit any conclusions to be drawn regarding nutrient losses in fruits stored in the home in the short interval between purchase and consumption. The results of various investigators cited by von Loesecke[2] indicated a lack of agreement with respect to ascorbic acid losses in stored tomatoes. Ascorbic acid was reported to diminish in stored apples and cranberries.

Considerable work has been done on ascorbic acid retention in frozen fruits and vegetables but most of it is primarily relevant to storage in commercial channels of distribution, rather than storage in the home freezer after purchase. Nevertheless, some useful data are available.

Dietrich et al[6] measured reduced ascorbic acid, dehydroascorbic acid and diketogulonic acid in peas, green beans and cauliflower held at temperatures ranging from 0° to 30° F. The rate of loss of reduced ascorbic acid increased with increasing storage temperature. In Laxton peas, for example, the loss of reduced ascorbic acid from about 15 mg/100 gm to about 2½ mg/100 gm, which occurred in one year at 10° F., took place in about 2 months at 20° F. and 2 weeks at 30° F. At 0° F. storage there remained after one year over 12 mg/100 gm. In peas and cauliflower stored at 20° F. there were decreases in total ascorbic acid, dehydroascorbic acid and diketogulonic acid while in green beans, even at 25° F. the total tended to remain constant.

Guadagni et al[7,8] carried out extensive studies on the effect of storage times and temperatures on frozen fruits, including strawberries, cherries, raspberries and peaches. In retail packs of frozen strawberries and red raspberries, Guadagni observed that there was little loss in the ascorbic acid content at 0° F. even after a year's storage, but rapid oxidation occurred at 10° F. In fact, ascorbic acid oxidation rates were found to increase logarithmically with increasing storage temperature. This relationship and the fact that the totals of reduced ascorbic acid, dehydroascorbic acid and diketogulonic acid remain fairly constant led Guadagni and Kelly[9] to suggest that these data can be used to determine the time-temperature exposure of a given sample. Further studies by Guadagni and Nimmo[10] on the effect of fluctuating temperature in the ranges of -10° to +10° F.; -5° to +5° F.; 0° to 20° F.; 10° to 30° F.; 15° to 25° F. on frozen raspberries and strawberries showed that it is the effective steady temperature (determined by the temperature coefficient in the range of the fluctuating cycle and the amplitude of the cycle) rather than the fluctuating temperature which relates to ascorbic acid losses. The temperature fluctuations to which a consumer may subject a package of frozen fruit by allowing it to remain at room temperature for an hour or more in a shopping cart, car and on a kitchen table before placing it in the ice cube compartment of a home refrigerator, might be comparable at least to the 0° to 20° F. cycle per 24 hours studied by Guadagni and Nimmo.[10] This cycle was found to result in a loss of 0.3-0.4 mg/100 gm of ascorbic acid per day in strawberries or raspberries. Byrne and Dykstra[11] quoted a study of temperature observations involving 16 frozen food holding-compartments in household type and refrigerator freezer combinations under a variety of operating conditions. Fifty-nine percent of the temperature readings fell between 0° and 20° F.; 11% registered above 20° F. Gordon and Noble[12] found no significant differences in the ascorbic acid content of blanched frozen asparagus and broccoli before and after 6 months of storage at 0° F., but they did find significant losses in the case of Brussels sprouts, cauliflower and green beans. However, when the ascorbic acid assays were carried out on samples after cooking, a significant difference relating to storage time was found only in the case of cauliflower. Wells et al [13] followed changes in the ascorbic acid content of several varieties of frozen green beans stored at O° F. for up to 9 months. The distribution of reduced ascorbic acid and its oxidation products in all samples remained practically constant throughout the storage period.

Payne [14] studied ascorbic acid changes in quick frozen and conventionally frozen corn stored at -30° F. for up to 12 weeks. In blanched, quick frozen corn cut off the cob there was no loss of ascorbic acid until after 3 weeks of storage and no difference between results obtained after 6 weeks or 12 weeks. Interestingly, increases in ascorbic acid were observed immediately after freezing and after certain periods of storage in corn that was quick frozen on the cob.

Eheart [15] studied the effects of refrigerated storage at 3° C. (38° F.) on fresh broccoli and of storage at -15° C. (5° F.) on the ascorbic acid content of frozen broccoli. Reduced ascorbic acid was not lost from broccoli stored 4 days at 38° F. Water-blanched frozen broccoli lost 31% reduced ascorbic acid on a wet basis between the first and fifth month of storage.

NUTRIENT LOSSES INCURRED IN PREPARATION

Nutrient losses incurred by fruits and vegetables in trimming, washing and soaking and chopping were reviewed by Harris.[2] He pointed out that in trimming, nutrient losses generally exceed weight loss because some nutrients are usually found in higher concentration in the outer leaves of vegetables and in the outer layers of tubers, roots and fruits. Ascorbic acid losses resulting from the peeling of potatoes ranged from 12-35%. Peeled potatoes stored under refrigeration for 20 hours lost little ascorbic acid and 8% thiamin. Potatoes soaked two hours in water lost 11.9% thiamin and sweet potatoes soaked five hours lost 21.1%.

In chopping, there is a loss of reduced ascorbic acid by oxidation to dehydroascorbic acid, which is about 80% as active biologically as reduced ascorbic acid. Harris quoted reports of losses of 19% and 52% of reduced ascorbic acid in the mincing and shredding of cabbage although corresponding losses in total ascorbic acid were 6% and 3% only. No further losses in reduced ascorbic acid occurred when shredded or minced cabbage was held for 3 hours. Other studies quoted by Harris reported losses of 9-15% in ascorbic acid, 0-5% in thiamin and 0-3% in riboflavin in diced cabbage, with no appreciable further loss when held at room temperature for 2 hours. Losses of 22% ascorbic acid were encountered in the slicing of cucumbers.

NUTRIENT LOSSES IN COOKING

Extensive studies have been carried out to ascertain nutrient losses in foods during cooking by various methods. An excellent review of these studies (through 1958) has been published.[2] In general, it was observed that when small portions of vegetables are boiled, nutrient losses vary according to the type of food, the stability of the nutrient, the amount of cooking water, the time of cooking and the cooking method used. In roots and tubers, thiamin retentions ranged from 67-100%, riboflavin retentions from 55-100%, niacin 39-100% and ascorbic acid 24-131%. In fruits, thiamin retentions ranged from 78 to 82%, riboflavin 92%, and ascorbic acid 32-76%. In herbage vegetables, thiamin retentions reported (when baking soda was not used) ranged from 47 to 100%; riboflavin from 55 to 100%, niacin 36-97%, and ascorbic acid 13-100%.

Hewston et al[1] studied the effects of boiling, covered or uncovered, on the retention of ascorbic acid, carotene, nicotinic acid, riboflavin, thiamin, calcium, phosphorus and iron in lima beans, green and white cabbage, corn, peas, green peppers, rutabagas, sweet potatoes, turnips and turnip greens. Ascorbic acid, thiamin and nicotinic acid losses were extensively studied in potatoes, carrots and peas. Ascorbic acid was the most sensitive of the nutrients studied with retentions generally in the order of 30-50%, although a loss of 18.3% was noted for 1″ green cabbage strips boiled uncovered for 60 minutes in 2.08 ml water/gm cabbage. A high of 99.4% retention of ascorbic acid was found in whole, unpared Katahdin potatoes boiled, covered for 40 minutes. Retentions were predictably lowest when the volume of cooking water was large, cooking time long and food particle size small.

Thiamin is both water soluble and heat labile and showed generally low retentions, which were related to the volume of water used. Except in cabbage, carotene was generally the most stable. Mineral losses were also related to the amount of water used in cooking.

Another important study of nutrient retention in vegetable cooking was the one carried out by Teply and Derse [16] on 24 frozen vegetable products which were subjected to cooking, using minimal amounts of water, to a subjectively determined "optimum flavor doneness." Leaching of nutrients was minimal except in the case of sodium, where 10-25% leaching occurred in most products, and in the case of turnip greens in which 15-25% of all vitamins except beta carotene were leached out. Many products retained 90% of vitamin C, but in cut snap beans, collards, corn and chopped spinach the loss ranged between 27 and 48%. In some vegetables, the destruction of beta carotene, folic acid, pantothenic acid and vitamin B_6 was as high as 50%. Considering over-all (i.e., solids plus liquor) recovery, there was essentially complete retention of thiamin, riboflavin and niacin.

Krehl and Winters [17] studied the effects of cooking methods on the retention of minerals (calcium, iron and phosphorus) and vitamins (thiamin, niacin, riboflavin, ascorbic acid and carotene) in 12 vegetables (asparagus, beets, broccoli, cabbage, carrots, cauliflower, corn, green beans, peas, potatoes, squash and spinach). In each case a 500 gm portion of raw vegetable was cooked by one of 4 methods: (1) pressure cooked with 125 ml (½ cup) of water, (2) boiled with sufficient water added just to cover the vegetable, (3) with 125 ml of water or (4) with no added water. The waterless cooking method resulted in the highest percentage of dry weight and the greatest retention of nutrients, while cooking with water to cover gave the lowest dry weight and the greatest nutrient loss. The lowest retentions (52.7-73.3%) were recorded for ascorbic acid, followed by thiamin (62.6-91.0%), niacin (59.1-91.3%) and riboflavin (63.1-87.9%). Higher retention levels were exhibited by carotene (79.1-94.7%) and the 3 minerals (72.1-94.4%).

Sweeney et al [18,19] investigated the effects of cooking methods on ascorbic acid and carotene retentions in fresh and frozen broccoli. Ascorbic acid retention in cooked fresh broccoli decreased as the quantity of cooking water increased, indicating that leaching was involved. This was confirmed by the fact that the total ascorbic acid losses (in solids plus liquids) were lower than those from solids alone. The transfer of ascorbic acid from broccoli solids to liquid was rapid (17-32%) in the first 5 minutes of cooking. Broccoli cooked to a satisfactory state of doneness retained from 60-85% of its original ascorbic acid in all cases except when an excessive amount of cooking liquid was used. Results obtained on frozen broccoli were similar to those obtained on fresh broccoli except that the absolute values were somewhat lower because of the lower initial ascorbic acid value of the frozen broccoli. No measurable loss of carotene content was noted in the broccoli when cooked by the methods investigated. Gordon and Noble [20] compared the effects of 4 cooking methods, (boiling water, tightly covered saucepan, pressure saucepan, and steamer) on ascorbic acid retentions in 11 vegetables. For the vegetables as a whole, the percentage retention in the boiling water method, 45% of the original content, was significantly smaller than in any of the steaming methods.

Cook and Sundaram [21] investigated the effects of boiling, steam-boiling (i.e., covered) and pressure cooking on the proximate composition, mineral content (Ca, P, Fe) and vitamin content (thiamin, riboflavin, niacin, panthothenic acid, biotin, vitamin B_6, folic acid, ascorbic acid and beta carotene), of artichokes. Significantly smaller amounts of phosphorus, iron, thiamin, riboflavin, niacin, pantothenic acid, vitamin B_6 and ascorbic acid were found in the boiled than in the raw artichokes. Except for beta carotene, ascorbic acid and vitamin B_6 the amounts of minerals and vitamins were higher in the steam-boiled and pressure-cooked buds than in those boiled in water to cover. The differences between the amounts of nutrients in the steam-boiled and the pressure-cooked samples were small and, in most cases, insignificant.

Bunnell et al [22] studied the alpha-tocopherol content of various foods and reported that home cooking of vegetables had but a minor destructive effect (average loss of 3.7% for 5 vegetables cooked in boiling water for ½ hour).

Sweeney and Marsh [23] studied the effects of cooking on stereoisomers of carotenes in fresh and frozen broccoli, Brussels sprouts, spinach, collards, kale, beet greens, endives, carrots, squash, red peppers and pumpkin, In the fresh and frozen green vegetables, cooking lowered the biologic value chiefly because of increased conversion of all-trans beta caratene (biopotency = 100) to neo-beta carotene U. (biopotency = 38). In fresh and frozen yellow and red vegetables, all-trans beta carotene was reduced by cooking with an accompanying increase in neo-alpha carotene B (biopotency = 16) and neo-beta carotene B (biopotency = 53).

A recent publication by Smith and Kramer [24] reported on losses in vitamins A and C in frozen collard greens cooked for varying lengths of time prior to freezing. The vitamin C content of the raw collard greens (109 mg/100 gm) was reduced to 31 mg/100 gm after 30 minutes cooking, 20 mg/100 gm after 60 minutes and 15 mg/100 gm after 2 hours. No decrease was found in beta carotene content.

There are relatively few reports in the literature on the effects of frying vegetables. Harris[2] cited a number of studies showing losses in ascorbic acid as a result of the home frying vegetables ranging from 0.7 to 80%. In one study, potatoes fried for 15 minutes retained 72% of their ascorbic acid, in another potatoes fried 20 minutes retained only 20-45%, and in still another 60% retention was found after 12 minutes.

Limited data are available also on the effects of cooking with additives. Hewston et al[1] studied the effect of added salt or soda to fresh peas and found that these had relatively little effect on ascorbic acid or nicotinic acid retention. Thiamin retention was actually slightly increased when soda or salt were used in cooking, but it should be noted that the addition of the soda reduced the required cooking time from 18 minutes to 8 minutes. Sweeney et al [18] also investigated the effects of various additives, including salt and sodium bicarbonate on the reduced ascorbic acid content of broccoli and found that the additives appeared to have little effect on ascorbic acid retention.

In recent years, microwave ovens have begun to make a serious entry in the field of domestic cooking appliances in the U.S. and consideration of the retention of nutrients attained in fresh or frozen vegetables cooked by this method is therefore becoming relevant to home use.

Thomas et al[25] compared conventional boiling, pressure cooking and baking with microwave cooking on the retention of carotene, thiamin, riboflavin and ascorbic acid in broccoli, cabbage, carrots and potatoes. Microwave cooking was not found to be better than boiling or pressure cooking for retaining ascorbic acid, carotene or riboflavin in vegetables. The effect of microwave cooking on ascorbic acid in broccoli and green beans was compared with that of the Chinese stir-fry method and conventional water cooking by Eheart and Gott.[26] Ascorbic acid retention for both vegetables was highest when they were cooked in a small amount of water (74-76%) but broccoli was found to retain as much ascorbic acid when stir-fried as when cooked in a small amount of water. Retention in microwave cooking (57-59%) was lower than in stir-frying broccoli but about the same as in stir-frying green beans.

These results are particularly interesting in the light of earlier data by Gordon and Noble[27] on cabbage, cauliflower, and broccoli cooked in boiling water, pressure saucepan or microwave oven, which showed that the mean percentages of ascorbic acid retained in samples cooked in the microwave oven (80-90%) were greater than those achieved in the pressure saucepan (70-82%). However, Kylen et al[28] have also reported that there were no statistically significant differences in the amounts of ascorbic acid retained in 7 fresh and 3 frozen vegetables cooked by a conventional and a microwave method.

LOSSES OF NUTRIENTS IN HOLDING PREPARED FOOD IN THE HOME

Data on the effects of holding prepared or cooked foods prior to service in the home are scarce. Harris[2] reported that orange juice held at 9° C. lost 17% thiamin in 24 hours, cantaloupe slices lost 35% ascorbic acid in 24 hours under refrigeration, and sliced bananas lost 12% ascorbic acid in 20 minutes at 25° C., while cucumber salad lost 8% ascorbic acid during 1 hour of standing.

Harris[2] reported a study showing that cooked broccoli, Brussels sprouts, shredded or sliced cabbage, cauliflower, peas and snap beans lost significant amounts of ascorbic acid during one day's refrigeration while cooked asparagus and spinach did not. Data on the extent to which cooked foods might be held hot in the home prior to service and the time-temperature relationship involved are not available, hence it would be improper to extrapolate from results which are available on the effects of steam table holding on nutrient losses. It should be pointed out, however, that the hot holding of many vegetables has been demonstrated to result in a lowering of ascorbic acid, thiamin and riboflavin levels.

CONCLUSIONS

It appears evident from the review that there are ample data availabe to begin to develop nutrient composition tables for vegetables "as eaten," at least with respect to ascorbic acid and, to a lesser extent, other vitamins and several minerals. More importantly, however, the data available should be used to convey to the consumer an understanding of ways of optimizing nutrient retention in the home preparation of fruits and vegetables.

References

1. Hewston, EM et al: Vitamin and mineral content of certain foods as affected by home preparation, Misc Pub No. 628, U S Department of Agriculture, 1948.
2. Harris, RS and von Loesecke, H: *Nutritional Evaluation of Food Processing*, New York: John Wiley & Sons, pp. 58-86; 418-435; 462-482, 1960.
3. Bring, SV et al: Total ascorbic acid in potatoes, *J Amer Dietet Assoc* 42:320-323, 1963.
4. Bring, SV and Raab, FP: Total ascorbic acid in potatoes, *J Amer Dietet Assoc* 45:149-152, 1964.

5. Page, E and Hanning, FM: Vitamin B$_6$ and niacin in potatoes, *J Amer Dietet Assoc* 42:42-45, 1963.
6. Dietrich, WC et al: The time-temperature tolerance of frozen foods. IV. Objective tests to measure adverse changes in frozen vegetables, *Food Tech* 11:109-113, 1957.
7. Guadagni, DG; Nimmo, CC; Jansen, EF: Time-temperature tolerance of frozen foods. VI. Retail packages of frozen strawberries, *Food Tech* 11:389-397, 1957.
8. Guadagni, DG: Nimmo, CC; Jansen, EF: Time-temperature tolerance of frozen foods. X. Retail packs of frozen red raspberries,*Food Tech* 11:633-637, 1957.
9. Guadagni, DG and Kelly, SH: Time-temperature tolerance of frozen foods. XIV. Ascorbic acid and its oxidation products as a measure of temperature history in frozen strawberries, *Food Tech* 12:645-647, 1958.
10. Guadagni, DG and Nimmo, CC: Time-tolerance of frozen foods. XIII. Effect of regularly fluctuating temperatures in retail packages of frozen strawberries and raspberries, *Food Tech* 12:306-310, 1958.
11. Byrne, CH and Dykstra, KG: Surveys of industry operating conditions and frozen product histories, *in* Van Arsdel, WB et al, *Quality and Stability of Frozen Foods,* New York: Wiley-Interscience, pp. 331-344, 1969.
12. Gordon, J and Noble I: Effects of blanching, freezing, freezing-storage and cooking on ascorbic acid retention in vegetables, *J Amer Dietet Assoc* 35:867-870, 1959.
13. Wells, CE et al: Ascorbic acid in uncooked frozen green beans, *J Amer Dietet Assoc* 43:559-561, 1963.
14. Payne, IR: Ascorbic acid retention in frozen corn, *J Amer Dietet Assoc* 51:344-348, 1967.
15. Eheart, MS: Effect of storage and other variables on composition of frozen broccoli, *Food Tech* 24:1009-1011, 1970.
16. Teply, LJ and Derse, PH: Nutrients in cooked frozen vegetables, *J Amer Dietet Assoc* 34:836-840, 1958.
17. Krehl, WA and Winters, RW: Effect of cooking methods on retention of vitamins and minerals in vegetables, *J Amer Dietet Assoc* 26:966-972, 1950.
18. Sweeney, JP et al: Effect of cooking methods on broccoli, *J Amer Dietet Assoc* 35:354-358, 1959.
19. Sweeney, JP et al: Palatability and nutritive value of frozen broccoli, *J Amer Dietet Assoc* 36:122-128, 1960.
20. Gordon, J and Noble, I: Effect of cooking method on vegetables, *J Amer Dietet Assoc* 35:578-581, 1959.
21. Cook, BB and Sundaram, S: Nutrients in raw vs cooked globe artichokes, *J Amer Dietet Assoc* 42:231-233, 1963.
22. Bunnel, RH et al: Alpha-tocopherol content of foods, *Amer J Clin Nutr* 17 (1):1-10, 1965.

23. Sweeney, JP and Marsh, AC: Effect of processing on provitamin A in vegetables, *J Amer Dietet Assoc* **59**:238-243, 1971.
24. Smith, JW and Kramer, A: Palatability and nutritive value of fresh, canned and frozen collard greens, *J Amer Soc Hort Sci* **97** (2):161-163, 1972.
25. Thomas, MH et al: Effect of electronic cooking on nutritive value of foods, *J Amer Dietet Assoc* **25**:39-44, 1949.
26. Eheart, MS and Gott, C: Chlorophyll, ascorbic acid and pH changes in green vegetables cooked by stir-fry, microwave and conventional methods and a comparison of chlorophyll methods, *Food Tech* **19**: 867-870, 1965.
27. Gordon, J and Noble, I: Comparison of electronic vs conventional cooking of vegetables, *J Amer Dietet Assoc* **35**:241-244, 1959.
28. Kylen, AM et al: Microwave Cooking of Vegetables - Ascorbic Acid Retention and Palatability, *J Amer Dietet Assoc* **39**:321-326, 1961.

Semiprocessed Fruit and Vegetable Products

Margaret Meinken, R.D.

HANDLING BEFORE PROCESSING

Fresh fruits and vegetables contribute texture, color, flavor and nutritive value to the American diet. The consumption of fresh fruits and vegetables has reached 55 billion pounds per year, or 261.9 pounds per capita assuming 210 million people as of July 1, 1972.[1]

Numerous authors agree that freshly harvested crops vary in their nutrient content. Fruits and vegetables are physiologically active after harvesting; enzymatic and respiratory activities continue with the crop. Preservation of freshly harvested fruits and vegetables is maintained through the normal channels of distribution by control of temperature. Proper control of temperature and humidity permits products to be held in the market prior to sale, preparation, and final consumption.

Before the introduction of hydrocooling, refrigeration or icing, a crop had limited distribution and economic value, since the harvested crop deteriorated rapidly. Today, through the wonders of hydrocooling and various rapid methods of transportation, fresh fruits and vegetables are available at any time during the year, benefiting both producer and consumer. Perhaps, in the future, a dip or chemical spray will evolve that will maintain fruits and vegetables in freshly harvested condition without need for humidity and temperature controls. Technology will advance through research.

In the respiration of fruits and vegetables oxygen is absorbed, carbon dioxide and water are given off; heat is produced. The amount of heat produced varies with each commodity and with

65

the temperature at which each product is stored (Table 1)[2]. Cooling by various methods reduces the heat of respiration and the resultant deterioration.

Ascorbic acid retention is frequently used as an index oi quality and nutritive value in fruits and vegetables. On the basis of extensive research, Fenton[3] drew the following conclusions: (a) "Of all the vitamins, vitamin C is most easily destroyed and no other vitamin or mineral is dissolved from vegetables more easily." (b) "If it is retained, all other attributes known as quality, that is aroma, color, flavor, texture, and nutrients, will also be retained."

Since ascorbic acid can be used as an index to quality and nutritive value, it is important to review some of its properties. Ascorbic acid, or vitamin C, is freely soluble in water, is odorless, and in pure form occurs as white crystals. Vitamin C is

Table 1.—Respiration of fruits and vegetables

Temperature and Product	Respiration
At. 60°F. Green Beans	10,802 B.t.u./ton/24 hr.
At. 32°F. Green Beans	1,694 B.t.u./ton/24 hr.
At. 60°F. Lettuce	22,600 B.t.u./ton/24 hr.
At. 32°F. Lettuce	638 B.t.u./ton/24 hr.
At. 80°F. Grapefruit	4,200 B.t.u./ton/24 hr.
At. 35°F. Grapefruit	750 B.t.u./ton/24 hr.
At. 80°F. Strawberries	42,000 B.t.u./ton/24 hr.
At. 35°F. Strawberries	3,300 B.t.u./ton/24 hr.

After McCoy, 1963[2]

rapidly oxidized to dehydroascorbic acid, which may be reversibly reduced to ascorbic acid. "If the pH of the medium is above 4, dehydroascorbic acid may be further and irreversible reduced to 2,3-diketogulonic acid and finally to oxalic acid and L-threonic acid."[4] Heat in the presence of air, alkali and catalysts such as iron and copper speeds the breakdown, while an acid medium and buffer substances tend to protect from oxidation.

"Values for this vitamin are reported most frequently on reduced ascorbic acid, the form in which most of the factor occurs in fresh foods. Dehydroascorbic acid, the oxidized form, is found in significant amounts in processed or stored foods. Since few investigators have reported both these forms, or total ascorbic acid, the data in tables of nutritive value may give lower than actual amounts."[4]

When green beans are picked on a warm day and stored overnight without refrigeration, they will lose 60% of their vitamin C content.[2] Likewise, ". . .cabbage, tomatoes and green peppers when kept for 48 hours in air at room temperature retained respectively 94, 88, and 85% of their original ascorbic acid - more than did kale (64%), carrots (56%), beets (75%), and peas (67%) . . . . Vegetables stored in crushed ice retained a greater percentage of ascorbic acid than did those kept on ice or in a refrigerator. Broccoli stored for 2 days in crushed ice had no ascorbic acid loss. When stored in the refrigerator, retention was 92%; on top of ice 91%; when left at room temperature, 56%."[4]

Losses of vitamins in fresh foods can occur through careless handling, but nutrient loss can be lessened by cooling, by avoiding bruising and by reducing the time between harvest and distribution. This applies to conservation of carotene as well as vitamin C. "Ezell and Wilcox found as much as 75% loss of carotene in kale and collards in 4 days at 70° F. under conditions of rapid wilting. At 50° F. there was 20% loss on slow wilting and 30% on rapid wilting. When wilting was prevented and kale stored at 32° F. there was a 25% loss of carotene in 4 weeks."[5]

Figure 1 shows the retention of ascorbic acid in raw green asparagus after holding at room temperature, and after cooling with chopped ice. Precooling and/or icing of raw asparagus as

Ascorbic Acid Content of Raw Green Asparagus Held at Room Temperature and Packed in Crushed Ice

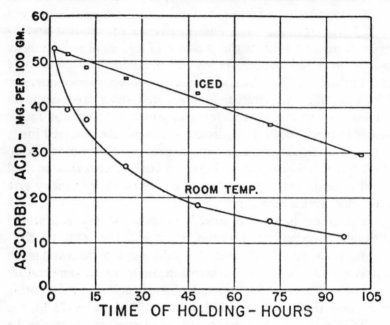

Figure 1   After National Canners, 1955[6]

soon as possible after harvesting is of great benefit in retaining the maximum amount of ascorbic acid.

Enzymatic and respiratory activities within the plant tissue can be reduced after the crop is harvested; through controlled conditions of holding, the quality and the nutritive value can be conserved.[6] Time and temperature are the key factors or essential elements in retention of nutrients within fresh produce.

PROCESSING

Segmenting and Juicing

The term "fresh" applied to fruit or vegetable juices means, to the Food and Drug Administration, that the product has not been processed in any manner. "Fresh" juice is freshly extracted in the retail establishment and handed to the customer without any processing.

Ascorbic acid in orange juice varies with the time the oranges are picked. Early season varieties as Parson Brown and Hamlin are highest, while late season Valencia variety is lowest.

The ascorbic acid loss during processing is negligible; the value for frozen reconstituted juice is the same as for Florida fresh juice.[7] Research done by the Florida Department of Citrus, and the University of Florida revealed that there is very limited detrimental effect on juices as a result of commercial juicing and processing. There is a small loss (perhaps 2%) of vitamin C as the juice is heat treated to reduce the activity of naturally occurring enzymes. From that point on, the vitamin C is relatively stable for reasonable periods of time.

Chaney[4] stated that more ascorbic acid will be secured from oranges or grapefruit which are halved, sliced, or sectioned than from those which are juiced. Certainly segmenting a fruit in a manner that does not break open cells would have less effect on vitamin C loss than a juicing operation which opens cells and exposes the vitamin to oxygen and enzymatic action.

Shredding

The ascorbic acid content of cabbage varies widely and is associated with variety, season, and climatic conditions of production. Vitamin C varies between heads produced under identical conditions and within heads.[8] This phenomenon is reported by many researchers. Wood et al[9] noted that the inner leaves of the cabbage head contained more ascorbic acid than either the middle leaves or the outer green leaves. However, Smith and Walker[10] found greater amounts of ascorbic acid in the outer green leaves of the cabbage head than in the inner white leaves. The method of cutting cabbage influenced the ascorbic acid retention.[9] Cabbage cut with a sharp knife retained slightly more ascorbic acid than cabbage cut by a dull knife or bruised by excessive pressure in shredding.

Numerous studies of the effect of holding shredded cabbage on ascorbic acid retention showed that little loss occurred under any of the methods studied. Stodola[11] evaluated the effects of three different methods of holding shredded cabbage: refrigerated and uncovered, refrigerated and covered with a damp towel, and shredded cabbage held at room temperature covered with a bag of ice. No significant loss of ascorbic acid occurred in the first two samples after 24 hours, nor in the third sample after six hours. An additional study was made on shredded cabbage held uncovered at room temperature. After holding six hours, the determinations showed not a loss, but a gain in ascorbic acid. This may denote the variation in shredded cabbage, or it may be the result of a loss of moisture in the cabbage, which resulted in a higher concentration of ascorbic acid.

Lampitt et al, [12] however, reported that ascorbic acid was lost rapidly during the first ten to fifteen minutes after cabbage was cut, then the value remained constant up to thirty hours, provided the temperature was approximately 59° F.

Cabbage used in my study [13] was not all from the same lot; since this study extended over a long period of time, individual units were obtained as needed. Reduced ascorbic acid was determined according to the method of Peterson and Strong. [14] Considerable variation was found in the ascorbic acid content within one wedge cut from a head of cabbage (Figure 2). The mean value of these determinations was 29.0 mg of reduced ascorbic acid/100gm of cabbage. Ascorbic acid content ranged from 19.4 mg to 43.8 mg per 100 gm of cabbage.

In another aspect of the study [13] the cabbage was shredded, mixed thoroughly, placed into plastic bags, and weighed. (To simulate adverse conditions which sometimes might occur, the cabbage was brought to room temperature, 70° F., before shredding.) Cooling rate and bacteriological studies were made of 41½, 10, 5, and 1 pound lots of the shredded cabbage, and some of the data collected are shown in Figures 3,4, and 5 and Tables 2 and 3. Figure 6 illustrates an effect of pre-cooling before shredding.

The experiment with the 41½ pound lot was discontinued at 72 hours because the temperature had not reached an acceptable level (as determined by thermocouples); the cabbage was discolored and had begun to decompose.

While it might seem logical to expect a loss in ascorbic acid on extended holding, the findings in this study corroborate finding of earlier studies. [11,12] Apparently, ascorbic acid is relatively stable in shredded cabbage held at refrigerator temperatures for extended periods of time.

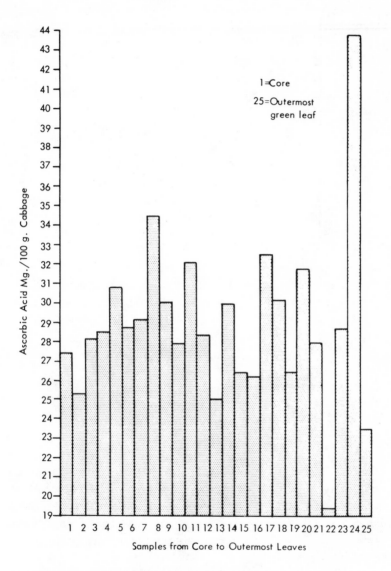

ASCORBIC ACID CONTENT OF DIFFERENT PARTS OF ONE HEAD
OF CABBAGE

Figure 2

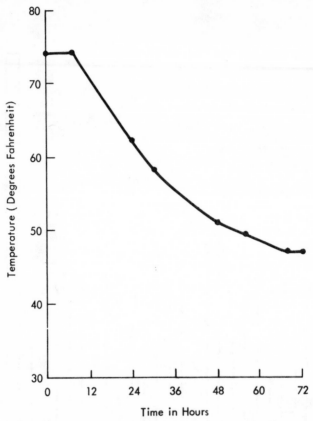

COOLING CURVES OF 41½ – POUND BAG OF SHREDDED
CABBAGE IN A REFRIGERATOR (38° F. ± 2°)

Figure 3

There was a marked increase in bacteria count after the second day of storage. Bacterial growth was determined by the method outlined in the *Recommended Methods for Microbiological Examination of Foods* by The American Public Health Association.[15] While the mean values were not always higher on successive days, a definite upward trend emerged. All of the counts were higher than those reported by Emery[16] and Fraier and Foster.[17]

Thus, the size of the package and the temperature of the product when processed have a definite relationship to the rate of cooling as well as to bacterial growth. The findings clearly

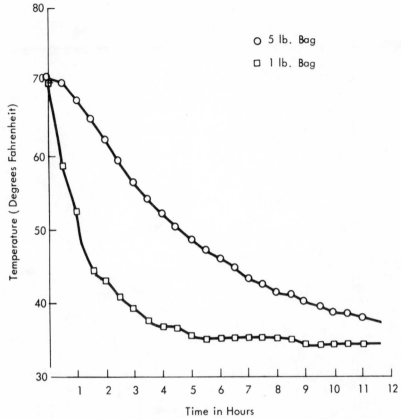

COOLING CURVES OF ONE AND FIVE POUND BAGS OF SHREDDED
CABBAGE IN A REFRIGERATOR (34° F. ± 2°)

Figure 4

indicate that small packages cool more rapidly than large ones. Ascorbic acid was very stable under the various conditions studied.

Since one cannot be sure that shredded cabbage which will be eaten raw has not been contaminated preceding or during processing, it is recommended that the temperature of the product be maintained below 45° F. throughout processing, storage and delivery. The essential elements of time and temperature should not be overlooked in the handling of any fresh fruit or vegetable.

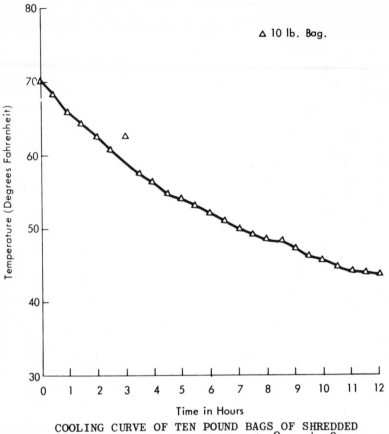

COOLING CURVE OF TEN POUND BAGS OF SHREDDED
CABBAGE IN A REFRIGERATOR (38° F. $\pm 2^\circ$)

Figure 5

Sulfite Dips

Traditionally, sulfur dioxide has been used to control undesirable browning reactions in drying fruits and vegetables, as a preservative, sanitizing agent, and as a bleaching agent.[18] Sulfur dioxide has been used widely in the wine industry for a long time to inhibit the growth of undesirable microorganisms in the fermentation process of grape juice.[19] The introduction of processed prepackaged vegetables such as potatoes has increased the use of sulfur dioxide and sulfites on food products.

The preservative quality of sulfur dioxide (sulfites, bisulfites, or metabisulfites), reduces, prevents or selectively inhibits spoil-

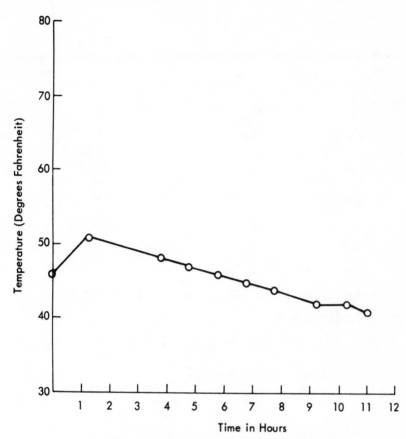

COOLING CURVE OF TEN POUND BAGS OF CABBAGE PRE-COOLED
BEFORE SHREDDING AND PLACED IN
A REFRIGERATOR (38° F. ± 2°)

Figure 6

age by microorganisms. [18,20] The antiseptic action of the chemical varies with microbial population, temperature, state of development, pH, and composition of the product being treated. [18] Frazier [19] notes that molds are affected more readily than yeasts or bacteria, when treated with sulfur dioxide.

Since sulfites are used widely in the food industry researchers have investigated the question of toxicity of this chemical to animals and to humans. Fitzhugh et al [21] found that sulfites were toxic to rats, when the sulfites were present in the diet in concentrations of 0.1% (615 ppm as SO_2) or more.

Table 2.—Ascorbic acid and bacteriological determinations on shredded
cabbage held in one 41½ pound plastic lined corrugated fiberboard box,
after 72 hours at 38° F. ± 2° (a)

Location of sample in box	Ascorbic acid mg./100 g. (b)	Bacteria/g. (c)
Top, center	18.0	450,000
Top, corner	26.2	400,000
Middle	19.5	189,900,000
Bottom	22.1	187,600,000
Mean of samples	21.45	94,587,500

(a) Zero time ascorbic acid 22.1 mg./100 g., bacteria
    70,000/g.
(b) Average of three titrations/sample
(c) Average of two samples

Cruess and Mackinney[22] found that carrots that were
blanched and treated with sulfites retained more carotene than
did those carrots that were unblanched and unsulfited. Smook
and Neunert[23] stated that sulfur dioxide treatment of apples
prior to preparation prevented the loss of ascorbic acid. If one
assumes these reports to be true, then dips in sulfite of mixed
vegetables for salads, to help retain their freshness, also would
tend to protect the carotene and ascorbic acid content.

CONCLUSION

Rapid handling and refrigeration of semiprocessed products
such as fresh vegetables and fruits are highly desirable. When the
raw vegetables are cut and partially processed, the microflora

Table 3.—Ascorbic acid and bacteria on shredded cabbage held in one pound plastic bags in a refrigerator (38° F. ± 2°) for two weeks.

Storage time in days	Ascorbic acid mg./100 g. (a)	Bacteria/g. (b)
0	22.1	680,000
1	25.7	650,000
2	24.2	630,000
3	29.0	1,360,000
4	25.9	1,540,000
5	25.0	2,020,000
6	19.9	2,040,000
7	25,8	4,660,000
8	23.0	3,560,000
9	17.6	1,290,000
10	27.3	4,100,000
11	25,7	---
12	25.8	---
13	21.6	12,800,000
14	24.7	9,800,000

Range 17.6 - 29.0

Mean 25.95

(a) Average of three titrations/sample.
(b) Average of three samples.

multiply rapidly after a few days. This research points out a definite need for further study to establish norms for total bacteria count and *E. coli* which could be considered safe for semiprocessed fruits and vegetables, especially those which will be eaten raw.

Recommendations based on research are needed for processing techniques, including methods for washing and sanitizing to reduce the microflora found on fresh fruits and vegetables as purchased. Recommendations are also needed for size of package, holding temperatures and safe time limits for storing the product before it is consumed.

A search of the literature confirms how little we know about semiprocessed foods and their nutritive values.

References

1. Magoon, CE: *Monthly Supply Letter*, United Fresh Fruit and Vegetable Association, July, 1972.
2. McCoy, DC: Refrigeration in Food Processing, in Joslyn, MA, Heid, JL (editors): *Food Processing Operations*, Westport, Conn: AVI Pub Co, Inc, 1963, vol I, p 372.
3. Fenton, F: Vitamin C Retention as a Criterion of Quality and Nutritive Value in Vegetables, *J Amer Dietet Assoc* 16:524-535, 1940.
4. Chaney, MS and Ross, ML: *Nutrition*, ed 8, Boston: Houghton Mifflin Co, 1971, pp 242-247.
5. Ezell, BD and Wilcox, MS: Loss of Carotene in Fresh Vegetables as Related to Wilting and Temperature, *J Agr Food Chem* 10:124, 1962, as cited in Bender, AE: Nutritional Effects of Food Processing, *J Food Tech* 1:261-289, 1966.
6. Cameron, EJ; Clifcorn, LE; Esty, JR; et al: *Retention of Nutrients During Canning*, National Canners Association, Washington, D.C., 1955, pp 51-56.
7. Watt, BK; Merrill, AL; Orr, ML: A Table of Food Values, in *Food. The Yearbook of Agriculture*, U.S. Department of Agriculture, Washington, D.C., 1959, p 232.
8. Bransion, HD; Roberts, JS; Cammeron, CR; et al: The Ascorbic Acid Content of Cabbage, *J Amer Dietet Assoc* 24:101-104, 1948.
9. Wood, MA; Collins, AR; Stodola, V; et al: Effects of Large-Scale Food Preparation on Vitamin Retention: Cabbage, *J Amer Dietet Assoc* 22:677-682, 1946.
10. Smith, FG and Walker, JC: Relation of Environmental Hereditary Factors to Ascorbic Acid in Cabbage, *Amer J Botany* 33:120-129, 1946.

11. Stodola, V: Ascorbic Acid Retention in Two Vegetables in Institution Food Service; Preliminary Preparation and Holding of Raw Potatoes and Raw Cabbage; Holding of Cooked Cabbage for Service. M.S. Thesis, Cornell University, 1945.
12. Lampitt, LH; Baker, LC; Parkinson, TL: Disappearance of the Ascorbic Acid in Raw Cabbage After Mincing and Chopping, *Nature* **149**: 697-698, 1942.
13. Meinken, ME: A Study of the Effects of Various Methods of Processing and Holding on Bacteria Counts, Ascorbic Acid Retention, and Percentage Waste of Selected Vegetables in Primary Processing at Central Food Stores, University of Missouri, M.S. Thesis, University of Missouri, Columbia, 1967.
14. Peterson, WH and Strong, FM: *Laboratory Manual for General Biochemistry*, Dubuque, Iowa: W.M.C. Brown Book Co, 1958, pp 83, 85.
15. *Recommended Methods for the Microbiological Examination of Foods*, American Public Health Association, Inc, 1958, p 99.
16. Emery, AW: How Detergents Improve Produce, *Food Eng* 31:57-59, *1959.*
17. Frazier, WC and Foster, EM: *Laboratory Manual for Food Microbiology*, Minneapolis: Burgess Pub Co, 1959, p 6.
18. Joslyn, MA and Braverman, JBS: The Chemistry and Technology of the Pretreatment and Preservation of Fruit and Vegetable Products with Sulfur Dioxide and Sulfites, in *Advances in Food Research*, New York: Academic Press Inc, 1954, vol V, pp 97-147.
19. Frazier, WC: *Food Microbiology*, New York: McGraw-Hill Book Co, Inc, 1958, p 134.
20. Deedes, F: Summary of Toxicity Data on Sulfur Dioxide, *Food Tech* **15**:28-33, 1961. p 622.
21. Fitzhugh, OG; Knudsen, LF; Nelson, AA: The Chronic Toxicity of Sulfites, *J Pharm Exp Therap* **86**:37-48, 1946.
22. Cruess, WV and Mackinney, G.: The Dehydration of Vegetables, Bulletin 680, University of California Agriculture Experiment Station, Sept 1943, as cited in Cruess, WV: *Commerical Fruit and Vegetable Products,* New York: McGraw-Hill Book Co, Inc, 1958, chap 19, p 622.
23. Smook, RM and Neunert, AM: *Apples and Apple Products,* New York: Interscience Pub, Inc, 1950, p 261.

Influence of Agronomic Practices on Nutritional Values

Milton Salomon, Ph.D.

When the term "agronomic practice" is limited to on-farm operations such as fertilizer application, liming, and use of pesticides and irrigation, the effects of such practices on mineral and vitamin content of fruits and vegetables are overshadowed and confounded by the nature of the soil, climate, time of planting and harvesting, and genetic selection. Additional variables of importance, such as storage, processing and handling, which affect nutritional quality of foods reaching the consumer, are not considered appropriate to this discussion.

Most of the research in the past has been concerned with yield of major food crops (wheat, corn, etc.) with less obvious interest in quality. Evidently it was assumed, and with some justification, that good, vigorous growth and high yields assured good nutritional characteristics. Also, estimates and methods for measuring nutritional quality were not well developed or understood. Greatest interest, in a nutritional sense, was in carbohydrates fat and protein with less concern for minerals and vitamins. There was early recognition of the strong influence and correlation of soils and climate with crop yield and composition.

Albrecht,[1,2] in the early 1940's, showed a clear relationship between the degree of soil development and plant composition. Plants from younger, less weathered soils of the western part of the country showed greater accumulation of potassium, calcium and phosphorus than plants grown in the East. Bear et al[3], conducting chemical studies on Rutgers tomatoes from 10 widely separated locations across the United States, reported remarkably wide differences in mineral element composition, i.e.,

81

magnesium, 5 to 59 me/100 gms; iron from 1 to 1938 ppm (!); and copper, from 0 to 53 ppm. Analyses of cabbage and snap beans showed similar wide variations. A compilation of ascorbic acid values[4] for spinach grown at various locations in New York State showed that the spinach grown on upland soils contained about 50% more ascorbic acid than that grown on muck soils (mean value 0.75 mg/gm as compared with 0.49 mg/gm). This was the case with several different varieties. The importance of location is further emphasized in a study reported by Reder,[5] in which ascorbic acid content of turnip greens was found to be more influenced by soil and climate than by nitrogen fertilization.

The effects of mineral fertilization, particularly on nitrogen distribution in vegetables has been well established.[6,7] However, the nature of major nitrogen reserves of nutritional consequence (protein) has been shown to be controlled more by genetic factors and species than by fertilizers and other cultural practices. Janes[8], in extensive fertilizer tests in Florida, showed that there were no significant differences in mineral composition (Ca, Mg, K, Mn, Fe and others) with different fertilizer rates up to 1½ times normal applications on cabbage, green beans, broccoli and collards. Uptake of major minerals and microelements when present in amounts adequate for normal plant growth is also related to distribution of associated elements in the soil, e.g., increase in pH with high rates of liming tends to depress the uptake of K and Mn in bean plants.[9] The effects of microelements of human and animal nutrition, including such often reported examples as iodine and cobalt deficiencies and selenium poisoning, are more related to the nature of the soil than to agronomic practices. Corrections can be made by additions to the soil through fertilizers or, as in the case of iodine, with supplements. Deficiency of boron is often associated with physiological disorders as root rots, stem cracking and water imbalance. Kelly[10] found a positive relationship between boron and carotene content of carrots. However, increase in carotene was more often associated with maturity. Boron appears to affect carbohydrate supplies rather than having a direct relationship with carotene in the root.

Recent concern about environmental poisoning of soil with

lead and other heavy metals has resulted in attempts to reduce plant uptake through changes in soil pH - a normal agricultural practice. John [11] reported a sharp reduction in lead accumulation by liming. Nitrogen fertilization, although it increased plant growth, did not appear to affect the uptake of lead. Nutritional or toxic effects of lead would appear to be of some concern when one considers the large and ubiquitous presence of this element in our atmosphere.

Recent studies on the mineral composition of "Eureka" lemons with potassium fertilization on deficient soils in California [12] show significant increases in several mineral elements and ascorbic acid content of lemon juice when potassium and phosphorus are applied separately at high rates. The results emphasize the common observation that complementary or competitive effects of fertilizer application on mineral uptake by the plant are strongly influenced by the nature of the soil, climate and plant species. The expectation of modifying nutritional values by fertilizer application must be considered in tandem with an appreciation of environmental conditions.

Some of the most productive vegetable areas in the country require irrigation. Janes [13] measured the mineral, ascorbic acid and carotene content of 2 varieties of snap beans and found very little effect on any of these nutritional factors by using irrigation rates ranging from none to heavy. In these studies chemical composition seemed strongly influenced by maturity of the plant which is also affected by water availability. Soil temperatures ranging from 60° F. to 75° F. Little difference in mineral composition, with the possible exception of calcium, was retures ranging from 60° F to 75° F. Little difference in mineral composition, with the possible exception of calcium, was reported. Composition varied more with harvest time than with temperatures.

It has long been recognized that the mode of action of some pesticides is closely associated with basic biologic processes of plant cells. It seems obvious that synthesis and metabolism of vitamins might also be affected by such chemicals. Several reports during the last few years indicate quite consistently that such may be the case. Recently, Sweeney [15] and Wu [16] have shown increases in beta carotene content of squash and carrots when treated with herbicides and soil fumigants. The differences in carotene levels between two squash varieties (Hubbard and Butternut) emphasize once again the influence of genetic variation on vitamin content. On the other hand, pre-emergent herbicides did not change the carotene content or macromineral composition of sweet potatoes in Kansas.[17] Soil fumigants (Telone, DBCP, EDB) used to control root rot nematodes of carrots shortened the time required for carrots to reach mature levels of beta-carotene.[18] Ascorbic acid content, so strongly influenced by light, did not change in the juice of tomatoes when soil were treated with several herbicides.[19]

For at least 20 years there has been a growing interest in foods produced by organic farming methods. In general, adherents to such practices claim that foods grown and marketed without the use of agricultural chemicals and in concert with companion cropping (polyculture) are nutritionally superior to those produced by conventional agronomic methods. Organic gardeners claim that soil organic matter enhances and changes the nutritional composition and quality of plant materials and is safer to use. This has not been proven and is open to question. However, organic gardeners are committed to heavy applications of manures and prepared composts as substitutes for chemical fertilizers. This does add a number of variables and dimensions to the problem which must be considered.

Unfortunately, and because of experimental complexities and other factors, the problem has not received adequate research. Little may be found in the literature specifically addressed to comparisons of the nutritional values of foods grown by the two systems. The problem is an important one, since the market for organic foods has grown remarkably and undoubtedly will continue to do so.

About the only significant and carefully researched reports published in this country comparing nutritional values of organically grown vegetables with those grown using chemical fertilizers have come from the U. S. Plant, Soil and Nutrition Laboratory at Ithaca, New York.[20] These findings indicate that there are no differences. In fact, the author, in summation, state that "no significant differences attributable to the source of plant nutrients were found." They also concluded that it seemed impractical to devise elaborate, time-consuming and expensive experiments for a more adequate evaluation of the problem until need for such work is clearly demonstrated.

With improved experimental techniques, greater understanding of nutritional significance of plant constituents, and greater attention to the problem, it is hoped that dependable data can be gathered in the future—data from which we can draw reliable conclusions.

References

1. Albrecht, WA: Calcium-potassium-phosphorus relation as a possible factor in ecological array of plants, *J Amer Soc Agron* 32:411-418, 1940.
2. Albrecht, WA: Potassium in the soil colloid complex and plant nutrition, *Soil Sci* 55:13-21, 1943.
3. Bear, FE; Toth, SJ; Prince, AL: Variation in mineral composition of vegetables, *Soil Sci Soc Amer Proc* 13:380-384, 1948.
4. Tressler, DK; Mack, GL; King, CG: Factors influencing the vitamin C content of vegetables, *Amer J Public Health* 26:905-909, 1936.
5. Reder, R; Ascham, L; Eheart, M: Effect of fertilizer and environment on the ascorbic acid content of turnip greens, *J Agr Res* 66:375-388, 1943.
6. Shannon, S; Becker, RF; Bourne, MC: The effect of nitrogen fertilization on yield, composition and quality of table beets (Beta vulgaris), *Proc Amer Soc Hort Sci* 90:201-208, 1967.
7. Lee, CY et al: Nitrate and nitrite in fresh, stored and processed table beets and spinach from different levels of field nitrogen fertilization, *J Sci Food Agr* 22:90-92, 1971.
8. Janes, BE: Composition of Florida-grown vegetables. III, Bulletin 488, Florida Agriculture Experiment Station, 1951.
9. Lynd, R and Turk, LM: Overliming injury on an acid sandy soil, *J Amer Soc Agron* 40:205-215, 1948.
10. Kelley, WC et al: Effect of boron on the growth and carotene content of carrots, *Amer Soc Hort Sci* 59:352-360, 1952.
11. John, M and Van Laerhoven, C: Effect of lime and nitrogen on lead content of lettuce, *J Env Qual* 1:169-171, 1972.
12. Embleton, TW; Jones, WW; Page, AL: Potassium and phosphorus effects on deficient "Eureka" lemon trees and some salinity problems, *Amer Soc Hort Sci* 91:120-127, 1967.
13. Janes, BE: The effect of varying amounts of irrigation on the composition of two varieties of snap beans, *Amer Soc Hort Sci* 51:457-462, 1948.

14. Singh, JN; Mack, HJ: Effects of soil temperature on growth, fruiting and mineral composition of snap beans, *Proc Amer Soc Hort Sci* **88**:378-382, 1966.
15. Sweeney, JP; Marsh AC: Effects of selected herbicides on provitamin C content of vegetables, *Agr Food Chem* **19**:854-856, 1971.
16. Wu, TC et al: Effects of certain soil fumigants on essential nutritive components and the respiratory rate of carrot (Daucus carota L.) roots, *Hort Sci* **5**:221-222, 1970.
17. Greig, JK and Al-Tikriti, AS: Effect of herbicides on some chemical components of sweet potato foliage and roots, *Proc Amer Soc Hort Sci* **88**:466-470, 1966.
18. Emerson, GA et al: Effects of soil fumigants on the quality and nutritive value of carrots, *Fed Proc* **30**:584, 1971.
19. Alban EK et al: Effect of selective herbicides on composition and quality of tomatoes and potatoes. Ohio Agr Research and Development Center Terminal Report, May 1965 - June 1969 (State Special Project 142).
20. Brandt, CS and Beeson, KC: Influence of organic fertilization on certain nutritive constituents of crops, *Soil Sci* **71**:449-454, 1951.

Varietal Influence on Nutritional Value

M. Allen Stevens

Traditionally plant breeders have been concerned with those characteristics that relate to yield - particularly disease resistance. Nutritional quality has not been a principal objective in most plant breeding programs. As a consequence, few varieties of fruits and vegetables have been developed for increased nutritional value and those that have been developed have usually been unsuccessful because of other weaknesses.

It is generally believed that fruits and vegetables are bought for flavor and appearance, i.e., for the enjoyment they provide, not for their nutritive value. Nutritional variation among varieties has undoubtedly not been considered by most consumers. How many consumers buy cos lettuce in preference to crisphead because the cos is more nutritious?

A number of factors contribute to the plant breeder's lack of attention to quality. Most of these are related to the complexity of breeding programs, which increases exponentially with the number of genes being manipulated. For example, in a cross segregating for 21 genes, a perfect population of tomatoes, which is one in which each phenotype occurs at least once, would require over 420,000 acres. Another factor is the lack of pressure on most plant breeders to hybridize and select for increased nutritional value.

To keep a breeding program manageable requires setting priorities. It is almost impossible to develop a variety with all of the desired characteristics. The best tasting, most nutritious variety is doomed to failure unless it has good yield, disease resistance, and the other characteristics acceptable to growers, shippers, and processors. In most breeding programs, quality is an adjunct, and often an afterthought. There is desire for as

87

much quality as possible, but not at the expense of yield or shipping and processing characteristics.

Fruits and vegetables that have the highest nutritional value are not usually the ones that contribute most to the nutritive needs of the population. Crops that are consumed in large quantities may make a significant contribution to nutrient intake even though their nutritional value is not relatively high. Obviously, if a crop is very low in a nutrient, high annual production of that crop contributes little nutritionally. For example, potatoes rank No. 1 in annual production of riboflavin; but a person would have to eat several pounds of potatoes per day to meet the Recommended Dietary Allowance for this nutrient. Obviously, rate of consumption has much to do with the nutritional contribution of a crop. In 1970, potatoes were the most prevalent vegetable, with over 16 million tons produced (Fig. 1). For comparison, about 7 million tons of oranges and 6 million tons of tomatoes were harvested.

United States Department of Agriculture Handbook No. 8[1]

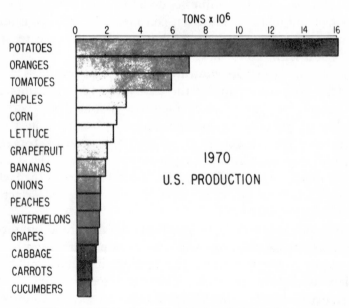

Figure 1.—1970 yield of the 15 most prevalent fruits and vegetables in the US.

and Agricultural Statistics for 1970[2] were used to estimate the relative nutritive value and relative contribution to nutrition of 39 major fruits and vegetables. Relative nutrition value relates to the composition of the crop, whereas relative contribution to nutrition considers both composition and annual production. The 39 crops were ranked for proximate concentrations of vitamins A and C, niacin, riboflavin, thiamin, potassium, phosphorus, calcium, iron, and sodium. The sum of the rankings for all of these components was used to estimate overall relative nutritive value. The product of nutrient concentration times 1970 yield was used to estimate relative contribution to nutrition. The sum of the rankings for relative contribution to nutrition for all of the above components estimates overall relative contribution to nutrition. This straightforward use of Handbook 8 and 1970 yield statistics imposes many limitations on the value of these estimates, but relative values are probably valid within broad limits.

A comparison of relative nutritive value and relative contribution to nutrition (Table 1) reinforces the often-heard lament that people are not eating the most of what is best for them. Generally, crops that are tops in nutritive value are not consumed in large quantities. Carrots and sweet potatoes are the only crops that are in the top 10 both for relative nutritive value and contribution to nutrition based on production.

The major nutritive contributions of fruits and vegetables are vitamins A and C. There are large intervarietal differences in these two components, and considerable research has been done on major crops with relatively high concentrations of these vitamins, particularly vitamin A.

VITAMIN A

As a result of their relationships to color, the components contributing to vitamin A content have received far more attention than have other nutrients. Carotenoids are vitamin A precursors, but to have vitamin A activity, a carotenoid must have at least one unsubstituted beta-ionone ring with a polyene side chain. Bauernfeind[3] has reviewed the vitamin A activity and occurrence of carotenoids.

Table 1.—Relative nutritional value and contribution to nutrition, based on production, of the top ten fruit and vegetable crops from each category.

	Rank for Relative Nutritional Value	Rank for Relative Contribution to Nutrition
Broccoli	1	21
Spinach	2	18
Brussels sprouts	3	34
Lima beans	4	23
Peas	5	15
Asparagus	6	25
Artichokes	7	36
Cauliflower	8	30
Sweet potatoes	9	10
Carrots	10	7
Corn	12	5
Potatoes	14	3
Cabbage	15	8
Tomatoes	16	1
Bananas	18	6
Lettuce	26	4
Onions	31	9
Oranges	33	2

Carrots rank at the top, both for vitamin A concentration and for contribution to consumption of this nutrient (Table 2). Other vegetables with high concentrations are sweet potatoes and spinach. Cantaloupes, apricots and broccoli are also good sources. In addition, tomatoes and peaches contribute importantly to vitamin A nutrition because of large production.

Table 2.—Vitamin A concentrations of several crops and their 1970 production of provitamin A. Included are the top ten crops in each category from among 39 major fruits and vegetables.

	IU/100 gm		Provitamin A (Tons - 1970)	
Carrots	11500	(1)	62	(1)
Sweet Potatoes	8800	(2)	37	(2)
Spinach	8100	(3)	9	(7)
Cantaloupe	3400	(4)	14	(4)
Apricots	2700	(5)	3	(11)
Broccoli	2500	(6)	2	(13)
Peaches	1330	(7)	12	(5)
Cherries	1000	(8)	2	(16)
Tomatoes	900	(9)	32	(3)
Asparagus	900	(9)	1	(21)
Peppers	420	(14)	5	(10)
Corn	400	(16)	6	(8)
Lettuce	330	(17)	5	(9)
Oranges	200	(21)	9	(6)

Carrots: There has been considerable research on the carotenes of carrots. Forty years ago Bills and McDonald[4] reported carotene concentrations ranging from 0.12 to 9.6 mg/100 gm among 10 varieties of widely different color. Harper and Zscheile[5] found total carotenoid concentrations ranging from 0.05 to 6.8 mg/100 gm among 16 commercial varieties and 18 imported lines, as well as large variations in the proportion of beta-carotene and alpha-carotene. Failure to consider variations in carotene ratios can result in considerable discrepancy between analytical and biological data for vitamin A activity. This is a failure in attempting to equate gross carotene concentration with vitamin A activity.

In a comprehensive study, Sadana and Ahmad[6] calculated vitamin A activity among red, orange, yellow and violet carrot varieties and found it to range from 30 to 19,500 IU/100 gm (Table 3). In the red varieties, beta-carotene was the principal pigment, and alpha-carotene and neo-beta-carotene B were also present in appreciable quantities. Alpha-carotene was the principal pigment in orange-colored varieties, and beta-carotene and xanthophyll were also present in relatively large quantities. Xanthophyll was the dominant pigment in yellow and violet varieties but only a small amount of beta-carotene was present.

Yamaguchi et al[7] found differences among four commonly used carrot varieties ranging from 9,900 to 15,000 IU/100 gm vitamin A. Their data agree closely with those of Howard et al[8] for the same four varieties. The most popular variety, Imperator, is a relatively rich source of vitamin A.

Sweet Potatoes: Sweet potatoes are also rich in vitamin A and make a large contribution to consumption of this nutrient (Table 2). Hernandez et al[9] have reported that sweet potato varieties vary greatly in carotene concentration (Table 4). The high carotene Centennial is the most popular variety today, and accounts for the majority of sweet potato acreage. Goldrush, which has an intermediate concentration, is also popular, but Georgia Red and Porto Rico are grown to a lesser extent. Howard et al[8] found large differences in vitamin A concentrations among the varieties, Velvet (14,000 IU/100 gm), Porto Rico (8,800 IU/100 gm), and Jersey (1,500 IU/100 gm).

Apricots: Apricots are a good source of vitamin A. Among eight varieties studied by Strachen et al [10] the concentration of carotene ranged from 1.21 to 3.49 mg/100 gm.

Peaches: In general peaches contain only about one-tenth as much provitamin A as do carrots. The concentration has been reported to vary from 0.4 to 1.2 mg/100 gm depending on variety.[10] The popular varieties, Elberta and Hale, are low in carotene

Tomatoes: Red-fruited tomato varieties generally have less than 10% of the vitamin A of commonly used carrot varieties. In a study of several species, varieties, and strains, Lincoln et al [11] found great variation in beta-carotene concentration. A 100-fold range was found in progeny from crosses between the

Table 3.—Concentrations of several carotenoids and calculated vitamin A activity of red, orange, yellow and violet carrot varieties.

Varietal Color	β Carotene	α Carotene	Neo-β-Carotene B (mg/100 gm)	Xanthophyl, γ, Neo-β-Carotene-U	Total Activity (IU/100gm)
Red	4.0 -9.6	1.0-1.6	0.5-2.0	0-0.3	8500-19500
Orange	.4 -1.6	1.3-2.6	---	.5-1.0	2200- 5200
Yellow	.02- .23	---	---	.2-1.5	30- 440
Violet	.02- .07	---	---	.2-0.7	30- 120

After Sadana & Ahmad[6]

Table 4.—Proximate carotene concentrations of several sweet potato varieties.

	β-Carotene (mg/100 gm)
Julian	18
Centennial	18
Goldrush	12
Heart O'Gold	6
Georgia Red	6
Kandee	6
Porto Rico	6
White Star	0
Pelican Processor	0

After Hernandez et al[9]

common tomato, *Lycopersicon esculentum* and another species, *L. hirsutum* (Table 5). In similar crosses between Baltimore and *L. hirsutum* and between Rutgers and *L. hirsutum*, Kohler et al[12] made a selection with an average concentration of 10.1 mg/100 gm of crude carotene, of which 88% was beta-carotene. This is about 20 times the beta-carotene in common varieties. From the latter cross, Tomes and Quackenbush[13] developed the high beta-carotene variety, CaroRed, which has about 10 times the beta-carotene of the recurrent parent, the commercial variety, Rutgers. The concentration of beta-carotene of Caro-Red is near the lower limits for common carrots. CaroRed is similar to Rutgers for many characters, except the high beta-carotene of CaroRed results in a distinctive red-orange color and different flavor. In a taste trial at the Indiana State Fair, with over 3000 individuals participating, Tomes and Quackenbush[13] showed that CaroRed has a flavor distinctively different from

Table 5.—β-carotene concentrations of tomato species, varieties, and strains.

Red & Yellow-Fruited Species	β-Carotene (mg/100gm)
L. esculentum	
Commercial varieties	.21 - .80
Baltimore (control variety)	.49 - .79
Foreign strains	.10 - 1.91
L. pimpinellifolium	.66 - 1.92
Green-Fruited Species	
L. peruvianum	.07 - .36
Cross of Red x Green	
F_1 (L. esculentum x L. hirsutum x L. esculentum)	.06 - 6.75

After Lincoln et al[11]

Rutgers. When color was masked, there was no flavor preference for CaroRed or Rutgers but when color was not masked, there was a preference for the red variety, indicating a color bias. Stevens [14] has also shown that carotenoid composition of tomatoes has an effect on flavor, apparently because several of the major volatile compounds of tomatoes result from enzymatic induced oxidation of the pigments. The oxidation is most common at the first conjugated diene bonds. From lycopene, the prevalent carotenoid of red tomatoes, the predominant compounds produced are 6-methyl-5-heptene-2-one and citral. From beta-carotene the principal compound is beta-ionone, the odor of which is readily apparent in high beta-carotene varieties.

Many tomato breeders are actively involved in attempts to incorporate the crimson and high-pigment characters, which influence vitamin A concentration, into successful tomato varieties. Compared to standard varieties, the crimson gene results in an approximately 50% decrease in beta-carotene, whereas high pigment results in about a 50% increase.

Corn: In 1919, Steenbock and Boutwell[15] discovered that yellow corn contained a fat-soluble vitamin needed to promote normal growth in rats, whereas white corn in the diet caused rats to die prematurely. During the 1920's, many investigators found that hogs gained weight faster with yellow corn than with white corn.

Quackenbush et al[16] reported that provitamin A in yellow corn inbreds ranged from a trace to 0.7 mg/100 gm: beta-carotene was the main provitamin A source.

In a diallel analysis of corn inbreds, Brunson and Quackenbush[17] found large variations in provitamin A among hybrids from various crosses indicating that potential for production of provitamin A is heritable. Crosses between inbreds with high provitamin A resulted in hybrids with high provitamin A concentrations, and vice versa. Hybrids from crosses between high and low concentration parents had intermediate concentrations. Data obtained by Quackenbush et al[18] indicate that corn inbreds vary greatly in the proportion of the three carotenoids that contribute to vitamin A activity (Table 6). Usually beta-carotene and cryptoxanthin are the most important contributors to vitamin A activity.

Oranges: Reported differences among orange varieties are not as great as the differences resulting from location.[19] Juice from Valencia oranges grown in California had about three times the vitamin A of Valencias grown in Florida.

Peppers: Some pepper varieties are a rich source of vitamin A. Howard et al[8] reported that red chili peppers contain 11,000 IU/100 gm vitamin A, whereas the popular green bell pepper contains only 500 IU/100 gm.

Lettuce: The popular crisphead lettuce varieties contain about 300 IU/100 gm vitamin A whereas the loosehead, green cos varieties contain about 2600 IU/100 gm.[8]

Watermelon: Tomes et al[20] found that beta-carotene ranged from 0.04 to 0.6 mg/100 gm among certain watermelon varieties (Table 7). The higher concentrations compare favorably with the provitamin A concentration of common red tomatoes.

Table 6.—β-carotene, β-zeacarotene and cryptoxanthin concentrations of several corn lines.

Line	β-Carotene (mg/100gm)	β-Zeacarotene (mg/100gm)	Cryptoxanthin (mg/100gm)
Os420	.27	.04	.14
H60	.18	.05	.31
D. C.	.20	.12	.21
Oh45	.44	.40	.32
Hy	.06	.01	.10

After Quackenbush et al[18]

Table 7.—β-carotene concentrations of several watermelon varieties.

Variety	β-Carotene (mg/100gm)
Charleston gray	.60
Mandella	.54
Candy red	.47
Strawberry	.36
Royal golden	.11
Purdue Hawkesbury	.04

After Tomes et al[20]

VITAMIN C

There are at least 18 major fruits and vegetables that can significantly contribute to vitamin C nutrition. Peppers, broccoli and Brussels sprouts are very rich sources of vitamin C, but oranges and potatoes make the greatest contribution to vitamin C nutrition (Table 8).

Peppers: Howard et al[8] found vitamin C concentration varied from 160 mg/100 gm for green bell to 245 mg/100 gm for green chili peppers.

Table 8.—Vitamin C concentrations and 1970 production of this vitamin by several crops. Included are the top ten crops in each category from among 39 major fruits and vegetables.

	mg/100gm	Tons (1970)
Peppers	128 (1)	471 (7)
Broccoli	113 (2)	342 (11)
Brussels sprouts	102 (3)	57 (28)
Cauliflower	78 (4)	184 (21)
Strawberries	59 (5)	292 (14)
Spinach	51 (6)	188 (20)
Oranges	50 (7)	7891 (1)
Cabbage	47 (8)	1116 (5)
Grapefruit	38 (9)	1434 (4)
Cantaloupe	33 (10)	443 (8)
Asparagus	33 (10)	85 (27)
Tomatoes	23 (16)	2729 (3)
Potatoes	20 (18)	6497 (2)
Corn	12 (20)	603 (6)
Bananas	10 (26)	360 (10)
Apples	7 (32)	436 (9)

Oranges: The reported range of intervarietal variation for vitamin C in the orange is not as great as in some other crops. Cohen [21] found a range from 51 to 78 mg/100 ml in juice. He also found that fruits from trees on sour orange rootstock consistently contained higher amounts of vitamin C than those on several other rootstocks. In contrast, Bitters [19] reported that the vitamin C concentration of Valencia oranges did not vary greatly on several rootstocks.

Tomatoes: In a study of 98 different varieties, MacLinn et al [22] found vitamin C concentration varied from 13 to 44 mg/100 gm. Shivrina [23] reported 100% difference between cultivated varieties and wild types. Lincoln et al [11] found vitamin C concentration ranged from 8 to 119 mg/100 gm among species, varieties, and strains studied (Table 9).

Varieties with double the usual vitamin C concentration have

Table 9.—Vitamin C concentrations of tomato species, varieties, and strains.

Red & Yellow-fruited species	Vitamin C (mg/100gm)
L. esculentum	
Commercial varieties	10.4 - 44.6
Baltimore (control variety)	19.3 - 31.6
Foreign strains	14.3 - 73.5
L. pimpinellifolium	40.1 - 86.5
Green-fruited species	
L. peruvianum	38.6 - 119.4
Cross of Red x Green	
F_1 (L. esculentum x L. hirsutum x L. esculentum)	8.4 - 43.4

After Lincoln et al [11]

been developed but they have not been commercially successful. There have been reports in the news media that the prevalent machine-harvest varieties have greatly reduced vitamin C concentrations. However, this is not true; the predominant ones have vitamin C levels comparable to those of hand-pick varieties, i.e., about 25 mg/100 gm.

Potatoes: In a cooperative study among several experiment stations, [24] intervarietal variation of about 100% was found. In Maine, the range was from 22 mg/100 gm (Golden) to 40 mg/100 gm (Potomac). In Minnesota the range was from 8 (Early Ohio) to 17 mg/100 gm (Mesaba). Relative varietal differences tended to be consistent among locations. The data obtained by Howard et al[8] indicate that the very popular Russet Burbank is a relatively poor source of vitamin C.

Allison and Driver [25] reported that variety is a significant factor in differences in the vitamin C concentration of potatoes and that these differences were effected by differential response of varieties to storage loss of vitamin C. Reaction to storage is crucial to vitamin C concentration as most potatoes are stored prior to consumption. Julén[26] found that the ascorbic acid concentration in some varieties was very high at harvest but diminished rapidly during storage. Other varieties had relatively low concentrations at harvest but only small losses during storage. Still others retained a high concentration during storage.

Squash: Howard et al[8] observed large differences in vitamin C concentration among squash varieties. Summer squash varied from 9 mg/100 gm (Zucchini) to 85 mg/100 gm (Bolsom pear). Variation among winter squash varieties was from 6 mg/100 gm (Pink Banana) to 21 mg/100 gm (Butternut).

Lettuce: Vitamin C concentration in lettuce varies markedly among head types.[8] Popular crisphead varieties have much lower concentrations (5 mg/100 gm) than do cos types (24 mg/100 gm).

Apples: In a study by Strachen et al [10] vitamin C concentration varied from 2 to 19 mg/100 gm among apple varieties (Table 10). Several high quality dessert apples - Delicious, Jonathan and especially McIntosh and Spartan - were low in vitamin C.

Table 10.—Vitamin C concentrations of several apple varieties.

	Vitamin C (mg/100gm)
Wegener	19
Northern spy	16
Rome beauty	11
Jubilee	10
Golden delicious	10
Winesap	9
Jonathan	7
Delicious (red strains)	6
Stayman	6
Spartan	3
McIntosh	2

After Strachen et al[11]

Peaches: Vitamin C is low in most peach varieties (5 mg/100 gm). An exception is Pavie Suncling which has 17 mg/100 gm, approaching the concentration in tomatoes. [27]

OTHER VITAMINS

Niacin: Among the 39 major fruits and vegetables, peas have the highest niacin concentration (Table 11). Potatoes, with about 50% of the niacin in peas, make the greatest contribution to niacin nutrition based on production. Corn, with about the same concentration as potatoes, also contributes significantly. Most vegetables and fruits cannot be considered good sources of niacin.

Riboflavin: Broccoli, spinach and asparagus are the best sources of riboflavin among major fruits and vegetables (Table 12). Potatoes produced the most riboflavin in 1970, but the

Table 11.—Niacin concentrations and 1970 production of this vitamin by several crops. Included are the top ten crops in each category from among 39 major fruits and vegetables.

	mg/100gm		Tons (1970)	
Peas	2.9	(1)	14.0	(6)
Corn	1.7	(2)	43.0	(2)
Potatoes	1.5	(3)	244.0	(1)
Asparagus	1.5	(3)	2.0	(20)
Lima beans	1.4	(5)	1.1	(25)
Peaches	1.0	(6)	15.0	(5)
Artichokes	1.0	(6)	0.3	(34)
Broccoli	0.9	(8)	1.3	(23)
Brussels sprouts	0.9	(8)	0.2	(35)
Cauliflower	0.7	(11)	0.7	(31)
Tomatoes	0.7	(11)	42.0	(3)
Bananas	0.7	(11)	13.0	(7)
Carrots	0.6	(16)	5.0	(10)
Watermelon	0.5	(22)	6.0	(9)
Oranges	0.4	(26)	32.0	(4)
Lettuce	0.3	(30)	7.0	(8)

amount is nutritionally insignificant because of the low concentration in the tuber.

Thiamin: Peas, lima beans, and corn, along with asparagus, are the best major vegetable sources of thiamin (Table 13). Variation from 0.07 to 0.16 mg/100 gm has been reported among potato varieties. [24]

MINERALS

Genetic variation influencing the concentration of certain minerals in some vegetables has been demonstrated, but there has been little study of this variation.

Table 12.—Riboflavin concentrations and 1970 production of this vitamin by several crops. Included are the top ten crops in each category from among 39 major fruits and vegetables.

	mg/100gm		Tons (1970)	
Broccoli	.23	(1)	0.3	(19)
Spinach	.20	(2)	0.4	(18)
Asparagus	.20	(3)	0.2	(21)
Brussels sprouts	.16	(4)	0.1	(33)
Peas	.14	(5)	0.7	(9)
Corn	.12	(6)	3.0	(2)
Lima beans	.12	(6)	0.2	(30)
Snap beans	.11	(8)	0.8	(7)
Cauliflower	.10	(9)	0.2	(28)
Peppers	.08	(10)	0.3	(26)
Lettuce	.06	(13)	1.4	(5)
Bananas	.06	(13)	1.1	(6)
Peaches	.05	(19)	0.7	(8)
Potatoes	.04	(26)	6.5	(1)
Tomatoes	.04	(26)	2.4	(3)
Oranges	.03	(32)	2.3	(4)
Apples	.02	(37)	0.6	(10)

Potassium: The predominant cation of fruits and vegetables is potassium (K). Lima beans and watermelons are the best sources among major horticultural crops (Table 14), but potatoes are by far the largest producers of K, based on annual production. The large production of oranges and tomatoes results in a relatively large contribution to K nutrition by these two crops. In a study of 55 tomato lines from divergent sources, Stevens[28] found K varied from 175 to 350 mg/100 gm in the fruits. The data indicated that the differences were under genetic control.

Table 13.—Thiamine concentrations and 1970 production of this vitamin by several crops. Included are the top ten crops in each category from among 39 major fruits and vegetables.

	mg/100gm		Tons (1970)	
Peas	.35	(1)	1.7	(5)
Lima beans	.24	(2)	0.2	(22)
Asparagus	.18	(3)	0.2	(20)
Corn	.15	(4)	3.8	(3)
Cauliflower	.11	(5)	0.1	(27)
Potatoes	.10	(8)	16.2	(1)
Watermelon	.10	(8)	1.3	(7)
Sweet Potatoes	.10	(8)	0.7	(11)
Spinach	.10	(8)	0.2	(23)
Broccoli	.10	(8)	0.2	(24)
Brussels sprouts	.10	(8)	0.04	(37)
Oranges	.09	(12)	7.1	(2)
Tomatoes	.06	(19)	3.2	(4)
Lettuce	.06	(19)	1.8	(6)
Bananas	.05	(24)	0.9	(9)
Grapefruit	.04	(28)	0.6	(10)
Apples	.03	(33)	0.9	(8)

Phosphorus: Lima beans, peas and corn are the best sources of phosphorus among major fruits and vegetables (Table 15). Potatoes, with about 35% the concentration in peas, produced the most phosphorus in 1970. In a study of the effects of phosphorus concentration on differential buffering in tomato fruits, Stevens[28] found that this element makes an important contribution to the aberrant relationship between total acid and

Table 14.—Potassium concentrations and 1970 production of this mineral by several crops. Included are the top ten crops in each category from among 39 major fruits and vegetables.

	mg/100gm		Tons (1970)	
Lima beans	650	(1)	503	(26)
Watermelon	600	(2)	8136	(4)
Spinach	470	(3)	865	(21)
Artichokes	430	(4)	145	(35)
Potatoes	407	(5)	66109	(1)
Brussels sprouts	390	(6)	110	(36)
Broccoli	382	(7)	579	(24)
Bananas	370	(8)	6655	(6)
Carrots	341	(9)	3074	(9)
Celery	341	(9)	2581	(12)
Corn	280	(16)	7033	(5)
Tomatoes	244	(20)	14477	(3)
Peaches	202	(26)	3042	(10)
Oranges	200	(27)	15781	(2)
Lettuce	175	(30)	4040	(7)
Apples	110	(38)	3423	(8)

pH and that phosphorus concentration is highly heritable. The concentration in a number of tomato lines varied from 20 to 83 mg/100 gm.

Calcium: Broccoli and spinach are good vegetable sources of calcium (Table 16). The concentrations among potato varieties vary considerably,[24] but generally potatoes contain only about 7% as much calcium as does broccoli.

Table 15.—Phosphorus concentrations and 1970 production of this mineral by several crops. Included are the top ten crops in each category from among 39 major fruits and vegetables.

	mg/100gm		Tons (1970)	
Lima beans	142	(1)	110	(21)
Peas	116	(2)	553	(6)
Corn	111	(3)	2788	(2)
Artichokes	88	(4)	30	(32)
Brussels sprouts	80	(5)	23	(35)
Broccoli	78	(6)	118	(20)
Watermelon	69	(7)	936	(5)
Asparagus	62	(8)	80	(24)
Cauliflower	56	(9)	66	(26)
Potatoes	53	(10)	8609	(1)
Onions	36	(15)	546	(7)
Cabbage	29	(17)	345	(10)
Tomatoes	27	(19)	1602	(3)
Bananas	26	(20)	468	(9)
Lettuce	22	(24)	508	(8)
Oranges	17	(31)	1342	(4)

Iron: Among major horticultural crops, spinach and lima beans are the best sources of iron (Table 17).

Rapid changes in technology and the continual appearance of new disease problems have kept plant breeders busy keeping the varieties available that are needed for an abundant supply of fruits and vegetables.

The recent Food and Drug Administration ruling requiring new varieties to meet certain nutritional and toxicant require-

Table 16.—Calcium concentrations and 1970 production of this mineral by several crops. Included are the top ten crops in each category from among 39 major fruits and vegetables.

	mg/100gm	Tons (1970)
Broccoli	103 (1)	156 (16)
Spinach	93 (2)	171 (15)
Snap beans	56 (3)	406 (7)
Lima beans	52 (4)	40 (26)
Artichokes	51 (5)	17 (33)
Cabbage	49 (6)	582 (4)
Tangerines	40 (7)	60 (22)
Celery	39 (8)	295 (10)
Carrots	37 (9)	334 (8)
Brussels sprouts	36 (10)	10 (38)
Onions	27 (13)	411 (6)
Lettuce	20 (14)	462 (5)
Grapefruit	16 (25)	302 (9)
Tomatoes	13 (30)	772 (3)
Oranges	11 (32)	868 (2)
Potatoes	7 (38)	1137 (1)

ments in order to be classified "Generally Recognized as Safe" means that breeders must become more actively concerned with the composition of the germ plasm and promising breeding lines.

There is undoubtedly much that breeders can do to improve the nutritional value of predominant varieties if and when this goal receives a top priority.

Table 17.—Iron concentrations and 1970 production of this mineral by several crops. Included are the top ten crops in each category from among 39 major fruits and vegetables.

	mg/100gm		Tons (1970)	
Spinach	3.1	(1)	57.0	(2)
Lima beans	2.8	(2)	2.2	(23)
Peas	1.9	(3)	9.0	(9)
Brussels sprouts	1.5	(4)	0.4	(35)
Artichokes	1.3	(5)	0.5	(34)
Broccoli	1.1	(6)	1.7	(24)
Cauliflower	1.1	(6)	1.3	(28)
Strawberries	1.0	(8)	2.5	(25)
Asparagus	1.0	(8)	1.3	(28)
Snap beans	0.8	(10)	5.8	(15)
Corn	0.7	(14)	18.0	(4)
Bananas	0.7	(14)	13.0	(6)
Potatoes	0.6	(19)	98.0	(1)
Watermelon	0.6	(19)	8.1	(10)
Tomatoes	0.5	(24)	30.0	(3)
Lettuce	0.5	(24)	12.0	(7)
Apples	0.3	(36)	9.4	(8)
Oranges	0.2	(39)	16.0	(5)

References

1. Watt, BK and Merrill, AL: *Composition of Foods, Agriculture Handbook 8,* U. S. Department of Agriculture, Washington, DC, 1963.
2. *Agricultural Statistics, 1971,* U.S. Department of Agriculture, Washington, D.C.
3. Bauernfeind, JC: Carotenoid vitamin A precursors and analogs in foods and feeds, *J Agr and Food Chem* **20**:456-473, 1972.

4. Bills, CE and McDonald, FG: The carotene content of ten varieties of carrots, *Science* 76:108, 1932.
5. Harper, RH and Zscheile, FP: Carotenoid content of carrot varieties and strains, *J Food Res* 10:84-97, 1945.
6. Sadana, JC and Ahmad, B: The carotenoid pigments and the vitamin A activity of Indian carrots, *Ind J Med Res* 35:81-91, 1947.
7. Yamaguchi, M; Robinson, B; MacGillivray, JH: Some horticultural aspects of the food value of carrots, *Proc Amer Soc Hort Sci* 60:351-358, 1952.
8. Howard, FD; MacGillivray, JH; Yamaguchi, M: Nutrient composition of fresh California-grown vegetables, bulletin 788, California Agricultural Experiment Station, 1962.
9. Hernandez, TP; Constantin, RJ; Miller, JC: Inheritance of and method of rating flesh color in *Ipomoea batatas*, *Proc Amer Soc Hort Sci* 87:387-390, 1965.
10. Strachen, CC et al: Chemical composition and nutritive value of British Columbia tree fruits, publ 862, Canada Department of Agriculture, 1951.
11. Lincoln, RE et al: Provitamin A and vitamin C in the genus *Lycopersicon*, *Bot Gaz* 105:113-115, 1943.
12. Kohler, GW et al: Selection and breeding for high beta-carotene content (provitamin A) in tomato, *Bot Gaz* 109:219-225, 1947.
13. Tomes, ML and Quackenbush, FW: Caro-red, a new pro-vitamin A rich tomato, *Econ Bot* 12:256-260, 1958.
14. Stevens, MA: Relationship between polyene-carotene content and volatile compound composition of tomatoes, *J Amer Soc Hort Sci* 95:461-464, 1970.
15. Steenbock, H and Boutwell, PW: Fat-soluble vitamine. III. The comparative nutritive value of white and yellow maizes, *J Biol Chem* 41:81-97, 1919.
16. Quackenbush, FW et al: Carotene, oil, and tocopherol content of corn inbreds, *Cereal Chem* 40:250-259, 1963.
17. Brunson, AM and Quackenbush, FW: Breeding corn with high provitamin A in the grain, *Crop Sci* 2:344-347, 1962.
18. Quackenbush, FW et al: Composition of corn: Analysis of carotenoids in corn grain, *J Agr and Food Chem* 9:132-135, 1961.
19. Bitters, WP: Physical characters and chemical composition as affected by scions and rootstocks, in *The Orange*, W.B. Sinclair (ed), Univ of California Division of Agricultural Sciences, 1961, pp 56-95.
20. Tomes, ML; Johnson, KW; Hess, M: The carotene pigment content of certain red fleshed watermelons, *Proc Amer Soc Hort Sci* 82:460-464, 1963.
21. Cohen, A: The effect of different factors on the ascorbic acid content in citrus fruits. II. The relationship between species and variety and the ascorbic acid content of the juice, Res Council Israel Sec D *Bot Bull* 5(2/3):181-188, 1956.

22. MacLinn, WA; Fellers, CR; Buck, RE: Tomato variety and strain differences in ascorbic acid (vitamin C) content, *Proc Amer Soc Hort Sci* **34**:543-552, 1937.

23. Shivrina, AN: A study of vitamins C and provitamin A (carotene) in tomato varieties, *Bull Appl Bot, Genet, and Plant Breeding,* suppl 84, Vitamin Prob, **2**:128-141, 1937, illus. [Eng summary in *Summaries,* p 6.] [*Chem Abs* **33**:1366, 1939.]

24. Leichsenring, JM et al: Factors influencing the nutritive value of potatoes, Bulletin 196, University of Minnesota Agriculture Experiment Station, 1950.

25. Allison, RM and Driver, CM: The effect of variety, storage and locality on the ascorbic acid content of the potato tuber, *J Sci Food Agr* **4**:386-396, 1953.

26. Julén, G: The potato as a source of vitamin C, *Lantbruks-Högsk Ann* **9**:294-309, 1941.

27. Souty, M: La vitamine C dans les peches destinees a la conserve, *Qual Plant et Mater Veget* **21**:223-228, 1972.

28. Stevens, MA: Relationships between components contributing to quality variation among tomato lines, *J Amer Soc Hort Sci* **97**:70-73, 1972.

The Influence of Harvest Time
on Nutritional Value

Tung-Ching Lee, Ph.D. and C.O. Chichester, Ph.D.

In order to simplify discussion, we will consider the term "harvest time" to represent a function of maturation (or maturity) of fruits and vegetables. The composition of fruits and vegetables is profoundly influenced by factors such as genetics, agronomic practices, regions and rate of growth, variety and climatic conditions. The difficulty of interpreting data in the literature in reference to maturity is further complicated by the different ways of sample collection, sample preparation, and sample analysis as discussed in other chapters of this book. For these reasons, it is not always possible to assign definite numerical values to composition changes during maturation. It is important to bear this in mind when interpreting the data given in the tables in this chapter.

Early man recognized that his survival depended on proper food choices based on appearance, odor, texture, and taste. Although an attractive food is not necessarily a nutritious food, the harvest time of fruits and vegetables does relate to organoleptic qualities. The best harvest time for some fruits appears to be a suitable stage of maturity for maximum transport and storage. For example, Hulme[1] reported that numerous attempts have been made to establish the point in time to harvest mangoes for shipment over short distances (short-term storage) or for intermediate length of storage. Other studies have been limited primarily to determining the best time for harvesting fruits or vegetables to meet a state or federal requirement. Furthermore, fruits and vegetables for canning purposes are usually picked when "firm ripe", ie, not quite ripe enough for table use, but of full size. Immature fruit is lacking in color and flavor, whereas overripe fruits and vegetables are apt to soften badly during processing in the can. In all these cases, obviously little attention has been given to the consideration of the nutritional quality of fruits and vegetables as it relates to the harvest time.

In spite of the general lack of a large amount of consistent data, some data are available in the literature to illustrate the relationship of maturity of fruits and vegetables to their nutritive values.

PROTEIN

The significance of fruits and vegetables as sources of protein is generally small compared with that of cereals and foods of animal origin.

Chaplin[2] conducted a comprehensive analysis of the amino acid composition of Stayman apple tissue in relation to maturity, ripening and senescence (Table 1). Although the total amino acid content increased slightly as the fruits matured on the tree, the profile remained constant.

In the case of tomato fruit, the total nitrogen content during ripening of fruit has been variously reported to decrease,[3,4] increase[5] or show no change.[6] Yu et al[7] showed that during the development of fruit the total nitrogen content fell from an initially high value in small green fruit to a minimum value near incipient ripeness, then rose again to a peak near the red stage and finally decreased towards overripeness. Yu et al also showed that the alcohol-soluble (non-protein) nitrogen increased with advancing maturity.

Hansen[8] commented that results from the investigations of protein content in fruit in relation to the fruit development and ripening established the dynamic nature of the changes taking place.

In vegetables, Whiteacre et al[9] showed there was an increase of protein content of turnip greens at successive stages of growth. Appleman and Miller[10] showed that similar results occurred in the white potato.

Table 1.—Total amino acid (free plus protein) composition of Stayman apple peel tissue in relation to harvest date.*

Amino Acid	5 Oct.	Harvest Date 12 Oct. (% of distribution)	19 Oct.	26 Oct.	Average
Aspartic acid	13.2	9.7	10.6	11.7	11.3
Threonine	6.6	5.3	7.4	6.8	6.5
Serine	10.1	7.0	10.2	9.1	9.1
Glutamic acid	6.3	5.8	5.6	7.3	6.2
Proline	6.0	6.2	7.3	6.7	6.6
Glycine	10.9	7.5	10.8	9.1	9.6
Alanine	12.2	10.5	11.0	11.1	11.2
Valine	7.5	6.3	7.6	7.6	7.2
Isoleucine	5.3	4.2	5.5	5.7	5.2
Leucine	8.4	6.6	9.3	9.0	8.2
Tyrosine	2.6	2.0	2.6	2.4	2.4
Phenylalanine	3.0	2.4	3.6	3.6	3.2
Lysine	4.8	4.0	4.6	4.8	4.6
Histidine	1.7	1.2	1.8	2.2	1.7
Arginine	1.4	2.0	2.1	2.6	2.0
		μmole per g dry weight			
Total amino acid content	77.2	100.7	105.9	100.7	

* from Chaplin[2]

CARBOHYDRATES

During the maturation of fruits and vegetables, the change in carbohydrates (sugar in its free forms or as its derivatives) is related to flavor, texture, appearance, and vitamin C content. Flavor is fundamentally the balance between sugar and acid, and flavor constituents which may be glucosides. The colors of many fruits relate to sugar derivatives of anthocyanidins. Tex-

tures of fruits and vegetables are governed by structural polysac-
charides. Vitamin C is a sugar derivative.

No consistent patterns emerge from studies of sucrose concentration during growth and maturation. For example, mango,[11] apricot, and peach[12] show a high sucrose content at
maturity. Tomatoes[13] have very little or no sucrose at any stage
of growth and maturation. The sucrose content of pineapple
increases during the last two months of maturation to a high
value.[14] Keitt[15] also showed different patterns of change in
sucrose concentration among different varieties of sweet potato.

Young fruits of apple, tomato, citrus fruits, mango and banana contain starch. In some fruit, the initial increase in the
concentration of starch is followed by a decrease, while in
others the concentration may increase up to maturity, eg, in
mango.[11] Vegetables also show different patterns of change.[15]

Reducing sugars often increase steadily throughout growth
and maturation in fruits such as tomato[13] and pineapple.[14]

Most changes in carbohydrate levels during maturation usually do not occur quickly or dramatically. Overmaturity causes
a decrease in carbohydrate content in some crops.

Although on a weight for weight basis, fruits and vegetables
in general are relatively poor sources of calories, the starchy
vegetables and legumes are staple foods in many areas and
contribute substantially to the total calorie intake—as much as
37%-40% of the total in some parts of Africa. Thus, harvest
time as it relates to carbohydrate content of vegetables in these
areas may have a real significance.

LIPIDS

The content of lipid materials in fruits and vegetables is
generally below 1%. However, olive, avocado, and palm accumulate oleaginous reserves during their development. Consequently, these fruits may represent an important source of fats for
human nutrition.

Leoncini and Rogai[16] studied changes in the composition of
olives at different stages of maturity, and found that oil and
sugars increased during the growth. Grado Cerezo[17] investigated the change in oil content during fruit development and

maturation, in an attempt to fix the best time for harvest and found that oil content increased with maturity.

Biale and Young[18] reported that Appleman showed a similar pattern of increased fat during the growth of avocado fruit.

VITAMINS

The contribution of fruits and vegetables to human nutrition lies in the area of vitamins and minerals.

Vitamin C (Ascorbic Acid) - Ascorbic acid is usually high in immature orange and grapefruit. As fruit ripens and increases in size, the concentration generally decreases.[19] However, when calculated on a per fruit basis, the total ascorbic acid usually increases.

Olliver[20] examined the ascorbic acid levels in developing black currants, gooseberries and strawberries. In black currants and gooseberries the ascorbic acid per berry increased very rapidly in the early stages and remained constant; thus as the fruit continued to increase in weight, there was a decrease per unit weight of fruit. In contrast to this, the curve of development of ascorbic acid in strawberries showed that the total ascorbic acid per berry remained low prior to color development when there was a rapid increase in ascorbic acid followed by a decrease at the end of the season.

Hobson and Davies[21] reported that the literature data on ascorbic content during the development and ripening of tomato fruit are inconsistent. They cited that some earlier investigators[22-24] reported little change, while more recent work has indicated an increase in ascorbic acid concentration during maturation,[25] with either a continuous rise[13,26] or a slight fall[27] during the final stages of ripening.

An increase in ascorbic acid during ripening of peppers[28] and snap beans[29] has been reported, with no change in the case of beets;[28] overmature peas were found to contain less ascorbic acid than the mature vegetable.[28] The ascorbic acid content of white potatoes has been reported by Lyons and Fellers[30] to remain rather constant during growth.

Provitamin A (Carotenoids) - The essential physiological function of carotenoids in animals is to act as precursors of vitamin

A. It is only in plants that carotenoid synthesis *de novo* has been shown. The provitamin A activity of carotenoids is based on the presence of an unsubstituted β-ionylidene ring (eg, in beta-carotene, alpha-carotene, gamma-carotene, and cryptoxanthin) and is quite specific for this group.

Provitamin A in carrots, peppers [28] and yellow corn [31] increases during maturation, but does not change during ripening of peas and beets. [28]

Khudairi [32] showed that during the ripening of tomato, chlorophyll peaks (430 and 662 nm in acetone-chloroform) disappeared and a prominent peak of phytofluene appeared at 365 nm. With further tomato color development, phytofluene was reduced and colored carotenes accumulated, absorbing at 445, 475, and 515 nm (Table 2). Hobson and Davies [21] reported a similar change based on data from Edwards and Reuter. [33]

Data are inconsistent concerning the changes in provitamin A during growth of sweet potatoes. Ezel et al [34] reported an increase in carotene during the first part of the season to a maximum concentration and then a decrease. Speirs et al, [35] however, reported no change in carotene during growth of the sweet potato.

Vitamins of the B group - Members of this group are all active prosthetic groups of tissue enzymes and are generally found in tissues which are metabolically active. Mature plant tissues do

Table 2.—Tomato discs of 1 cm diameter were extracted with acetone-chloroform (1:1) and measured spectrophotometrically.*

Tomato ripening stage	% absorbance at nm					
	365	430	445	475	515	662
Mature green	34	40	-	-	-	30
White (turning)	98	-	38	42	36	-
Orange red	53	-	48	57	50	-
Red	31	-	53	64	57	-

* from Khudairi, A.K. [32]

not normally show high rates of metabolic activity and are relatively poor sources of the B vitamins. Rather higher levels are found in the meristematic tissues of actively growing shoots and in the embryos of seeds. The cereals are the main plant sources of these vitamins, however legumes and other vegetables may supply folic acid, thiamin, niacin, riboflavin, pantothenic acid, biotin and vitamin B_6.

Peynand and Ribereau-Gayon [36] reported that generally the total vitamin content of grapes (Merlot) increases as the fruit matures, with the exception of biotin content which decreases. Apart from nicotinamide and mesoinositol, which are still increasing at harvest, the vitamin content reaches its peak some days or weeks before the fruit is completely ripe.

In the case of vegetables, thiamin and niacin increase during maturation of snap beans, asparagus, [29] and spinach, [37] and decrease during ripening of lima beans. [38] An increase in riboflavin content of snap beans during growth has been reported by Flynn et al. [29]

MINERALS

Fruits and vegetables contain wide ranges of mineral elements and make an important contribution to human nutrition in this area. In contrast to the extensive literature on mineral contents of vegetables and fruits as related to the composition of soil, fertilizer practices and other agricultural factors, the studies concerning the changes in the mineral components during the maturation and ripening were relatively few.

There is generally an increase in ash content and thus a corresponding increase of minerals during the growth of vegetables. Flynn et al [29] reported an increase in calcium and phosphorus during maturation of snap beans. Whiteacre et al[9] also showed the increase in minerals as turnip greens mature. Peynand and Ribereau [36] reported that the cation content of the Merlot grape increased 2 or 3 times in the skin during maturation.

During the same period, heavy metals increase by 50%. There is also an increase in anions, particularly phosphate, which is concentrated in the seeds. The phosphate content also increases steadily in the peel and pulp as the fruit ripens.

SUMMARY

Because of the relative unavailability of data, the inconsistency of some of the available data and the complicated nature of research in this field, it is clear that there is a need for additional basic research.

We desperately need a standardization of terminology of maturity such as that proposed by Gortner et al.[39] They describe a biochemical basis for horticultural terminology of development, maturation, ripening and senescence. The basis for the proposal is a series of physical and biochemical changes occurring in a very reproducible manner and at specific times during fruit development.

More uniform sampling procedures, better analytical methodology, and systems of statistical data analyses will have to be applied before we can evaluate the relationship between the nutritional quality of fruits and vegetables and their harvest times.

Interdisciplinary approaches utilizing plant physiologists, biochemists, nutritionists, pomologists, and geneticists, backed by support programs are also essential if data in this field are to be developed.

References

1. Hulme, AC: in *The Biochemistry of Fruits and Their Products*, Hulme, AC (ed), New York: Academic Press, 1971, vol 2, chap 1.
2. Chaplin, MH: Amino Acid Changes in Stayman Apples after Harvest, M. Sc. Thesis, Rutgers University, New Brunswick, New Jersey, 1966.
3. Sando, CE: Bulletin 859, US Department of Agriculture, 1920.
4. Neubert, P: *Arch Gartenb* 7:29, 1959.
5. Rosa, JT: *Proc Amer Soc Hort Sci* 22:215, 1925.
6. Vendilo, GG: Vest mosk gos univ ser VI, *Biol Pachv* 19(2):52, 1964.
7. Yu, MH; Olson, LE; Salunkhe, DK: *Phytochem* 6:1457, 1967.
8. Hansen, E: in *The Biochemistry of Fruits and Their Products*, Hulme, AC (ed), New York: Academic Press, 1971, vol 1, chap 6.

9. Whiteacre, J; Fraps, GS; Yarnell, SH; Oberg, AG: *Food Res* **9**:42, 1944, as reported in Harris, RS and von Loesecke, H: *Nutritional Evaluation of Food Processing,* New York: John Wiley & Sons, 1960, chap 2, p 69.

10. Appleman, CO and Miller, EV: *J Agr Res* **33**:569, 1926.

11. Biale, JB: *Adv Food Res* **10**:293, 1960.

12. Deshpande, PB and Salunkhe, DK: *Food Tech* **18**:1195, 1964.

13. Dalal, KB; Salunkhe, DK; Boe, AA; Olson, LE: *J Food Sci* **30**:504, 1965.

14. Singleton, VL and Gortner, WA: *J Food Sci* **30**:19, 1965.

15. Keitt, TE: Bulletin 156, South Carolina Agriculture Experiment Station, 1911, as reported in Harris, RS and von Loesecke, H: *Nutritional Evaluation of Food Processing,* New York: John Wiley & Sons, 1960, chap 2, p 72.

16. Leoncini, G and Rogai, F: *Bull Inst Sup Agr* **(Pisa)** **8**:711, 1934.

17. Grado Cerezo, A de: *Bol Inst Invest Agr* **8**:101, 1943.

18. Biale, JB and Young, RE: in *The Biochemistry of Fruits and Their Products,* Hulme, AC (ed), New York: Academic Press, 1971, vol 2, chap 1.

19. Harding, PL and Fisher, DF: Tech'Bulletin 886, US Department of Agriculture, 1945, p 1001.

20. Olliver, M: *Analyst* **63**:2, 1938.

21. Hobson, GE and Davies, JN: in *The Biochemistry of Fruits and Their Products,* Hulme, AC (ed), New York: Academic Press, 1971, vol 2, chap 13.

22. Maclinn, WR and Fellers, CR: Bulletin 354, Massachusetts Agriculture Experiment Station, 1938.

23. Wokes, F and Organ, JG: *Biochem J* **37**:259, 1943.

24. Kaski, IJ; Webster, GL; Kirch, ER: *Food Res* **9**:386, 1944.

25. Georgiev, HP and Balzer, I: *Arch Gartenb* **10**:398, 1962.

26. Fryer, HC; Ascham, L; Cardwell AB; et al: *Proc Amer Soc Hort Sci* **64**:360, 1954.

27. Dalal, KB; Salunkhe, DK; Olson, LE: *J Food Sci* **31**:461, 1966.

28. Pepkowitz, LP; Larson, RE; Gardner, J; Owens, J: *Plant Physiol* **19**:615, 1944.

29. Flynn, LM; Hibbard, AD; Hogan, AG: *J Amer Dietet Assoc* **22**:413, 1946.

30. Lyons, ME and Fellers, CR: *Amer Potato J* **16**:169, 1939.

31. Scott, GC and Belkengren, RO: *Food Res* **9**:371, 1944.

32. Khudairi, AK: *American Scientist* **60**:696, 1972.

33. Edwards, RA and Reuter, FH: *Food Tech* (Aust.) **19**:352, 1967.

34. Ezell, BD; Wilcox, MS; Crowder, JN: *Plant Physiol* **27**:355, 1952.

35. Speirs, M; Peterson, WJ; Reder, R; et al: Bulletin 30, Southern Coop. Serv., 1953.

36. Peynand, E and Ribereau-Gayon, P: in *The Biochemistry of Fruits*

and Their Products, Hulme, AC (ed), New York: Academic Press, 1971, vol 2, chap 4.

37. Gleim, EG; Tressler, DK; Fenton, F: *Food Res* 9:471, 1944.

38. Eheart, JE; Moore, RC; Speirs, M; et al: Bulletin 5, Southern Coop. Serv., 1946, p 24.

39. Gortner, WA; Dull, GG; Krauss, B: *Hort Sci* 2:141, 1967.

The Technology of Handling Fresh Fruits and Vegetables*

Dale L. Anderson, Ph. D.

Foods that are dried, processed, canned, frozen or preserved by other methods have provided the means for the development of specialized societies, with many people concentrated in urban areas supplied with food by a commercial agriculture. However, people still prefer some of their fruits and vegetables in fresh form. These fresh items are now grown in specialized areas, and most are delivered tremendous distances to our heavily concentrated urban population against some very substantial obstacles. My purpose here is to provide a somewhat synopsized description of that complex marketing process.

Fresh fruits and vegetables are live, growing products during the marketing cycle; that is, respiration continues after harvest. However, the growing process of fresh fruits and vegetables changes when they are harvested, as there are no longer any nutritive or moisture inputs from the plant. Thus, the handling of fresh fruits and vegetables (and some plant materials) involves a set of conditions different from those for other perishable products. In addition, fruits and vegetables are not uniform in their requirements. Many different sets of conditions must be met for individual products, seasons, and areas, although we are often able to group like-products for handling purposes.

HARVEST

The harvesting of these products creates the first of several abrupt changes in their environment on the way to market.

* Official information material, U.S. Government; this chapter may be reproduced in part or in full.

121

"Hand picked" products are rare today. Economic pressures and labor shortages have moved producers toward mechanized systems such as mechanical harvesting, bulk handling, and continuous-line processing. Plant breeders have developed products more resistant to rough handling, various treatments, and longer storage and transit periods.

Harvest involves some method of collecting the product, such as picking, cutting, digging, or shaking - with the product placed, thrown, or dumped into some container. This could be a field crate, pallet bin, hopper, truck, or wagon. Products are collected at intervals and shipped to a processing or packing plant. Some crops are still harvested by a "mule train" moving through the fields, with harvested vegetables placed directly on the packing line belt. The packed product comes out the back of the train and is loaded for moving to a vacuum cooler or shipping dock. The current trend, however, seems to be away from field packing and toward mechanical harvesting, bulk handling, and fixed location packing sheds. Rough handling most frequently occurs when the fruit is separated or when it is dumped into the holding medium.

COOLING

High ambient temperatures, normal for most products at harvest time, can be determined once the product is removed from the plant. While considerable attention is paid to rapid removal of this field heat, the goal is not always accomplished.

Several methods are used to bring products to the desired temperature levels for marketing. Some products are held in cold storage after packing, which permits a rather slow removal of heat. Other products may be hydrocooled or vacuum cooled to speed up this process. In some instances, removal of most field heat may be left to the transit vehicle itself, which may or may not be capable of providing the refrigeration service.

The products vary, of course, in their temperature requirements. Such items as fresh sweet corn and iceberg lettuce should be cooled as rapidly as possible to 32° to 34° F. Other products are susceptible to chill damage at much higher temperatures or are damaged by rapid chilling. Early Florida grapefruit should be held at a temperature of 60° F.; whereas late grape-fruit should be held at 50° to 55° F.

PACKING OR PROCESSING

Normal processing procedures start with the product in temporary storage indoors. The product may be dumped onto a line, sorted, washed, trimmed, graded, waxed, or given other appropriate treatment. After these processes are completed, the product is packed, precooled, and sent back to storage or shipped. This series of handling steps may vary for different products, but few items escape some such complex and often rough process.

The increasing tendency toward high-volume, continuous-line processes increases the risk of damage and cross-contamination. With high-speed lines and rapid cooling, produce is often bounced, dropped, crushed, and subjected to pressures, scuffs, nicks, bruises, and chill damage that may not show up until later. Washing, hydrocooling, and waterflow conveying (as with potatoes) may provide opportunities for cross-contamination if the water baths used are not properly cleaned, treated, and controlled.

A variety of chemical, heat, or other treatments is used to control market diseases and market quality. Products may be waxed, chemically treated, or given a heat or chill bath during processing. Some treatments are used during packaging. Even those products packed in shipping containers may be hydrocooled or vacuum cooled or placed in storage with controlled or gaseous atmospheres. Unfortunately, a treatment that may solve one problem can also cause others. The whole marketing process involves a game of identifying and resolving new and old problems, both of quality and economics. USDA handbooks and other publications by the score help to provide up-to-date information in solving problems for growers, handlers, shippers, and receivers of fresh produce.

SHIPPING CONTAINERS AND PALLETS

Packaging constitutes a considerable cost to growers and shippers. The traditional shipping container, which has the

packer's brand beautifully displayed on the end and the produce neatly hand-packed on the top layer, was designed for
auction market sales where the product had to sell itself. Modern marketing methods, however, usually bypass this marketing
stage. Products are sold on the basis of grade or reputation, and
often the buyer never sees the product. Some products are
shipped on pallets which may be sent to retail outlets in pallet
units. However, most shipping containers are subject to many
individual handlings on the way to market. In highly automated
packing-houses, handling of shipping containers is completely
mechanical up to the railcar or truck, at which point the
containers are manually loaded. Palletized transportation systems are needed and are being adopted for some products.

The high costs of good packaging materials may result in the
use, by some shippers, of lower cost materials that provide
inadequate protection. Sometimes trade practices, such as over-
filling, result in product damage. Problems can also be caused
by the use of vibration devices to fill boxes instead of some of
the costly, manual labor.

The need for standardization of shipping containers is serious
in the fresh produce industry. Many types, sizes, and shapes are
currently in use for many of the same products, and most do
not palletize well. In one survey, it was found that 371 sizes
were used for 49 produce commodities.

A growing problem with shipping containers for produce
today is what to do with the waste material at the retail store.
Such materials as wirebound crates, wax-treated fiberboard,
plastic liners, and produce trimmings are difficult to recycle and
constitute a considerable disposal problem.

PREPACKAGING

We now estimate that over half of all fresh produce is pack-
aged in consumer units at some point in the system. Packaging
within a store, however, is expensive for the retailer. The trend
toward mechanization, and the opportunity it provides to pre-
pack and leave the inedible portions of fresh produce in the
producing areas, causes continuous pressure on grower/shippers

to prepackage. However, they have difficulty in collecting the full costs of prepackaging from the buyers, even though most produce reaches the consumer in a better, fresher package when it is prepackaged close to the point of sale.

Some products are shipped in bulk refrigerated hopper cars (160,000 pounds per car) or in bulk bins to packing plants in areas near the point of sale. Potatoes, onions, radishes, citrus fruits, and even bagged carrots have been shipped this way. These products are handled by conveyors and bulk handling systems and are prepackaged in consumer units. In some instances, returnable shipping containers are used to ship packages to retail outlets, thus eliminating the problem of packaging material disposal at the retail store. However, such shipping containers are presently used for an insignificant part of fresh produce shipments.

LONG DISTANCE TRANSPORT

After the packaging problems come the transportation problems. Long distance transportation for fresh produce has shifted heavily from rail to truck. At the same time, an increased specialization in production areas and concentrations of consumers farther from production have resulted in larger hauls and longer hauls. An increase in import/export business has also increased the transit distance for many fresh products.

Temperature control in transit depends primarily upon mechanical refrigeration, which, in transit trailers, is accomplished by blowing cold air in one end of the trailer and circulating it through the load. With meats or frozen food, it is necessary to cool only the perimeter of the load and to remove the heat that enters the vehicle through the walls, ceiling, or floor. However, with fresh produce, it is also necessary to cool the interior of the load, where heat is generated by product respiration. The problem is that the circulating cold air in the trailer does not always go where it is supposed to—some of it follows the path of least resistance and returns to the starting point without going through or around the load. The result is hot or cold spots and sometimes freezing of portions of low temperature loads. To relieve this problem, loading patterns have been developed to provide circulation channels through the load and to force the cool air to go through these channels. However, the use of proper loading patterns depends on human skills. In addition, these loading patterns waste transportation space.

Some loads are not properly precooled, which places extra demands on the refrigeration system of the transit vehicle. Table 1 gives desirable transit temperatures for selected fresh fruits and vegetables, shows the variations between products, and gives an idea of the problems faced with mixed loads.

Even though products may be at the proper temperature upon reaching their destination, they may well have been above the desired temperature for a considerable period of time before arrival; for example, if any stops for vehicle or refrigeration maintenance were necessary during transit. Good and rapid maintenance services are not always readily available away from home base.

The calibration of the thermostat that controls the refrigeration system of the vehicle can be affected by vibration and shocks during transit. When the load has been improperly braced, these vibrations and shocks can also cause crushed or damaged products from high-stack loading or load shifting. We do not know at this time how much effect such repeated vibration has on physiological aspects of the product.

Researchers are trying, with some degree of success, to make transportation vehicles and refrigeration more foolproof. However, loss and damage claims generally discourage transportation companies from seeking perishable cargo. In addition, many carriers are not purchasing adequate refrigerated equipment to meet the demand for refrigerated transportation. Specialized carriers, often owned by the grower/ shipper, are handling more perishable cargo each year.

With the increased interest in the export of fruits and vegetables, it is essential that improvements be made in transportation equipment and schedules. Research in this area is meeting with some success, and when good results are obtained in overseas shipments, the findings will be helpful in domestic transportation. However, foreign growers are also watching this research

Table 1.—Desirable transit temperatures.

Product	Temperature(°F.)
Apples	32-40
Avocados	
Most varieties	45
West Indian varieties	55
Bananas (green)	56-60
Cherries (sweet)	32
Cranberries	36-40
Dates	40-50
Grapefruit	50-60
Limes	48-50
Oranges	
Arizona & California	40-44
Florida & Texas	32-40
Asparagus	32-36
Beans (snap)	45
Cantaloupe	35-40
Celery	32
Cucumbers	45-50
Honeydew melon	45-50
Lettuce	32
Onions (dry)	32-40
Peppers (sweet)	45-50
Potatoes	
Early crop	50-60
Late Crop	40-50

and may use the results to send their products into the U.S. market. The export/import trade also increases disease and insect control problems and poses problems for exporters who must meet a wide variety of restrictions imposed by various governments.

Air transportation has not, as yet, gained a significant share of the fresh produce movement. It is used largely for small-lot or highly perishable but high-priced cargo. The problems with air transportation usually occur on the ground when perishable products are left in the hot sun or are improperly handled.

TERMINAL MARKETS

At the end of the long haul, fresh produce encounters a new host of problems.

Refrigerated cars may be left on team tracks or in yards and used as cold storage until the market needs the product. From this time on, temperature variations, rough handling, and unsanitary conditions frequently occur during unloading, drayage, warehousing, and order selection.

When the produce reaches the market centers, many of which are still antiquated, inefficient, and unsanitary, it often is placed on wet, muddy sidewalks or concrete floors. Even in new city wholesale markets, modern warehouses, and retail outlets, the sanitary conditions for handling produce often leave much to be desired. In general, the sanitary conditions we know in other fields are not yet found in the fresh produce industry.

Some produce may be handled on pallets or other unit handling systems in the warehouse and in delivery to large supermarkets. Slow moving products or products going to smaller retailers or restaurants may go through a repacking procedure. In these instances, lots are broken up for distribution. Once a shipping container is opened and the product removed, protection becomes a problem. Produce seems to deteriorate much more rapidly once it is given this extra exposure.

Warehouses may have two or three temperatures in storage zones; for instance, a wet box, a dry box, and general warehouse space which may or may not be air conditioned. Some may have banana ripening rooms. Most storage temperatures from this point on are determined by compromise.

DELIVERY

Produce is frequently not protected adequately during its delivery from the warehouse to the retail store or restaurant. Delivery vehicles often carry mixed loads of products which require a variety of temperatures; usually, the temperature used in such loads is a "reasonable" average. In some instances, produce is delivered in unrefrigerated delivery vehicles.

RETAIL

Although some retail stores and restaurants may have good receiving operations in which the produce is placed under refrigeration quickly, many do not. In addition, the temperature of the refrigerated storage in most retail outlets is not maintained at a low enough level to provide extended quality protection. These outlets rely on quick movement of the product to prevent its deterioration. However, control by the stores of the length of time that produce is held in their refrigerated storage before it is displayed for sale often leaves much to be desired. In addition, display cases, most of which are refrigerated, constitute short-term protection at best.

There was a time when retail produce clerks used a variety of washing, soaking, and trimming methods to return vegetables to a fresh appearance. Today, the clerks do not have the time or training to perform such functions; however, they do trim and package some of the produce, and some produce not sold during the day is reworked.

Produce does go through a number of handlings, both by clerks preparing and building displays and by customers during selection. Once selected, the product moves in a shopping cart to checkout and is placed in a grocery bag for the customer to take home.

When the produce reaches its final destination—the home of the consumer—it may have little freshness left, even though it appeared acceptable when it was purchased. The produce that the housewife must then discard, shortly after taking it home, is a result of the rapid temperature variations and rough handling it received during the various stages of its journey.

If we pick the product green enough, use a variety that is tough enough, and do not mistreat it early in the marketing system, we may have a product that will be accepted, with some satisfaction, by consumers. Those of us who know what fresh fruits and vegetables taste like, especially tree-ripened fruit, still seek something better than we are getting. Often, we try to buy good produce from farm markets. We must admit, however, that the fresh fruit and vegetable industry has made tremendous strides in raising the quality of mass-produced products.

THE CONSUMER

After the consumer has completed the grocery shopping and the purchases are carefully packed into shopping bags, the groceries are taken directly (?) to the home refrigerator. Of course, the best intentions are to shop last, but occasionally a beauty parlor appointment or an impromptu errand must be taken care of before the groceries are taken home. Next, the bags of groceries may rest in the kitchen for a time while the refrigerator is prepared to receive additional items. Bits of leftovers that have to be cleared from the refrigerator always include a variety of produce items—such as lettuce nubs, radishes, and carrots that have deteriorated beyond the edible stage.

Sometimes the best and most appropriate attention is given to the newly purchased produce. At other times, the fruits and vegetables are hastily placed in the vegetable crispers until a later time when they can be given further attention.

There are probably few generalizations that can be made concerning temperatures in the average home refrigerator. Table 2 shows the situation in one 12-year old refrigerator (my own) in which the air temperature in the front jumped from 42° F. to 70° F. within 30 seconds when the door was opened.

Table 2.—Temperature checks on a 12-year-old, large, upright Westinghouse refrigerator half full of food, with the control set on position 3 of a 5-position control dial, and with a room temperature of 76° F.*

Door position	Location of Temperature Measurements in Refrigerator		
	Top two-thirds 2 in. from door	2 in. from back wall	In closed veg. crisper
	°F.	°F.	°F.
Door closed			
20 min.	42	40	42
Door wide open			
15 sec.	55	53	42
Door open 90°			
30 sec.	70	62	42
1 min.	70	67	47
2 min.	72	69	45
3 min.	73	69	44
Door closed			
15 sec.	68	62	
30 sec.	58	56	
1 min.	53	51	
2 min.	50	48	
3 min.	47	78	
4 min.	46	46	
5 min.	45	45	
6 min.	45	45	

*Air temperature in refrigerator freezer, 16° F.

CONCLUSIONS

In summary, we have followed our fresh fruits and vegetables to market over a rough and sometimes rocky road. In general, they reach their destination in reasonably good shape. In the future, the fresh fruit and vegetable industry will face additional problems. These include competition from imports, high labor costs (somewhat offset by mass production techniques that, in turn, may cause other problems), longer transport distances and time, and a lack of public understanding of modern production and marketing methods.

On the positive side, I foresee better varieties, larger, more efficient production and packing operations, better packaging and precooling, better transport equipment, more modern marketing facilities, and better retail refrigeration. In particular, I foresee more prepackaging and more prepreparation of fresh ready-to-cook or ready-to-serve produce.

The Influence of Storage, Transportation and Marketing Conditions on Composition and Nutritional Values of Fruits and Vegetables*

P. H. Heinze, Ph.D.

Fruits and vegetables supply a large portion of a number of vitamins and minerals in our diet. They may also supply significant quantities of the major components such as carbohydrates and, in the case of vegetables, protein. The changes and losses in some of these constituents in fresh fruits and vegetables after harvest may be important if the movement of the commodity from the producer through the transport, storage, and marketing chain to the consumer is unduly long or if the conditions during this period are unfavorable for the commodity.

Fresh fruits and vegetables are living tissues when harvested and retain vital functions until senescence and decay or processing stop them. Enzyme activity plays a major role in any changes in composition in these tissues and is greatly influenced by the temperature and humidity conditions to which the fruits and vegetables are subjected. Thus, temperature and humidity are two big factors to be considered in any after harvest handling of fruits and vegetables. Obviously any handling implies careful avoidance of mechanical damage to the commodity.

---

* Official information material, U.S. Government; this chapter may be reproduced in part or in full.

The period of transport may be regarded as a portion of the storage period. The wholesaling, retailing, and the holding periods in consumers homes are variable and difficult to standardize in experimental tests. Some storage studies have included a final holding stage at conditions differing from those of the main storage period in an attempt to simulate the effects of the last steps in the marketing chain.

Many fruits and vegetables ripen during storage. Ripening is a general phenomenon not directly dependent upon storage of the commodity. Storage conditions primarily affect the rate of ripening; therefore, the influence of storage on ripening and the many changes associated with ripening are not discussed in detail in this paper.

Storage conditions may have a pronounced effect on the vitamin content, on some of the major constituents such as carbohydrates, and may cause some compounds to be converted into toxic materials. Newer storage techniques such as the use of controlled atmospheres may, under certain conditions, have an influence on the composition of fresh fruits and vegetables. A discussion of these topics will form the remainder of this presentation.

Ascorbic Acid

Vitamin C or ascorbic acid is one of the more important nutrients supplied by some fresh fruits and vegetables and is also one of the most sensitive to destruction when the commodity is subjected to adverse handling and storage conditions. Plant tissues contain oxidase systems that are capable of oxidizing ascorbic acid. Unfavorable conditions involving high or low (nonfreezing but chilling) temperatures, physical damage and wilting will all produce stress conditions in the tissues and accelerate the oxidation of ascorbic acid.

Leafy vegetables usually keep best when stored just above their freezing temperatures. These low temperatures are also most effective in maintaining vitamin C content. Kale and other leafy vegetables are very subject to wilting. The loss of ascorbic acid in kale has been found to be closely associated with the degree of wilting (Table 1). The same degree of wilting causes

Table 1.—Effects of Temperature and Wilting on Loss of Ascorbic Acid in Kale During 2 Days of Storage. (After Ezell and Wilcox[1])

Rate of Wilting	Average loss (percent)		
	32°F.	50°F.	70°F
Slow	2.4	15.3	60.9
Moderate	3.8	15.8	69.6
Rapid	5.3	33.1	88.8

much higher losses at 50° F. and 70° F. than at 32° F.[1] Other leafy vegetables such as spinach, turnip greens, and collards respond similarly. Cabbage, also a leafy vegetable, loses ascorbic acid more slowly than some of the previously mentioned vegetables. Whether this is because of its structure or because of a protective mechanism is not clear.

Low temperatures are not conducive to maintaining ascorbic acid content in all fruits and vegetables. Some are injured at temperatures well above their freezing temperatures. These commodities, including sweet potatoes, cucumbers, squash, tomatoes, bananas, and others, are termed "chilling susceptible." Sweet potatoes stores at 15° C. for 10 weeks were found to retain over 75 percent of their ascorbic acid content. Comparable roots stored at 7.5° C. lost 90 percent of their ascorbic acid during the same period. The decrease in ascorbic acid was accompanied by a 500 percent increase in chlorogenic acid.[2]

Potatoes, although relatively low in ascorbic acid content, have been considered a rather important source of vitamin C in the diet because of their extensive consumption. This significance has undoubtedly diminished with the increase in the more highly processed forms of potatoes such as chips and french fries. Many investigators have found that ascorbic acid content of fresh potatoes decreases during storage and that the loss is greater as the storage temperature is reduced below 50° F. (Table 2).[3] Potatoes stored at 40° F. have been found to lose as

Table 2.—Losses of Ascorbic Acid from Potatoes Stored for 7 Months at Different Temperatures. (After Murphy[3])

Variety		Average loss (percent)			
		32°	36°	50°	70°
Sebago	- 1st yr.	74	73	51	52
	- 2nd yr.	76	75	58	63
Irish Cobbler	- 1st yr.	67	69	53	43
	- 2nd yr.	76	78	55	55
Katahdin	- 1st yr.	65	72	52	50
	- 2nd yr.	65	67	51	58

much ascorbic acid in 2 months as comparable lots stored at 50° F. or 60° F. in 5 months.[4] In some instances storage at near freezing temperatures has produced rather striking temporary increases in ascorbic acid content for 1 or 2 months but this is followed by a rapid decline to levels much below that in potatoes stored at higher temperatures.

The loss of ascorbic acid in peas and beans may be retarded by storing these vegetables in the pods. Shelled lima beans lose ascorbic acid at twice the rate of unshelled beans at the same temperature (Table 3).[5]

Many fruits, particularly citrus, are good sources of vitamin C. There are conflicting reports on the stability of ascorbic acid in citrus. It is generally agreed that lemons retain nearly 100 percent of their ascorbic acid during storage and that grapefruits also lose very little ascorbic acid. Reports differ for oranges and tangerines but it has been concluded that loss of vitamin C in fresh citrus is unlikely to exceed 10 percent under reasonable conditions of distribution and marketing immediately after harvest.[6] Many fruits and fruit products are used in the fresh or fresh-frozen form, thus most of their ascorbic acid content is directly available and not partially destroyed in a cooking process as is more customary for vegetables.

Tomatoes retain ascorbic acid well during storage and may

Table 3.—Effect of Storage on Ascorbic Acid Content of Lima Beans. (After Eheart et al[5])

		Average loss (percent)	
Storage		Room Temperature	Refrig- erator
Shelled	- 2 days	67	16
Unshelled	- 2 days	39	5
Unshelled	- 4 days	72	19

Average content of freshly shelled beans was 85 to 97 mg per 100 grams dry weight.

show a slight increase with red color development. In general, there is a modest decrease during storage. Firm ripe tomatoes held for 10 days at 35° F or 50° F lost about 25 percent of their ascorbic acid.

Folic Acid

Folic acid is also lost to a considerable extent if storage conditions are unfavorable. The legume seeds, asparagus, spinach, turnip greens, and some other leafy vegetables supply significant quantities of folic acid to the diet. These vegetables stored for 2 weeks at refrigerator temperatures or in crushed ice were found to lose little or none of their folic acid content.[7] However, storage at room temperature for 3 days resulted in losses of 50 to 70 percent of the vitamin (Table 4).

Thiamin and Riboflavin

Fruits and vegetables are not considered important sources of thiamin and riboflavin but some of the legume seeds may supply significant amounts of thiamin. Fresh green shelled lima beans stored at room temperature for 2 to 4 days retained 90

Table 4.—Folic Acid-Content of Vegetables Under Various Storage Conditions, (gamma/gram dry basis). After Fager et al[7]

Vegetable	Fresh	Room temperature 3 days	Refrigerated 2 weeks	Crushed Ice 2 weeks
Asparagus	12.00	3.20	8.40	14.00
Endive	6.30	1.80	6.50	7.60
Green beans	5.00	2.20	2.80*	3.40*
Leaf lettuce	10.00	9.00	15.00*	20.00*
Parsley	12.00	7.00	10.00	14.00
Peas	0.97	0.56	0.89	1.08
Spinach	27.00	16.00	27.00	31.00
Swiss chard	7.30	6.30	7.00	9.70

*1 instead of 2 weeks

percent of their thiamin and nearly an equal percent of riboflavin (Table 5). Thus, these vitamins are relatively stable under unfavorable commodity storage conditions. Tests with green beans, carrots, corn, peas, and spinach indicate that thiamin and riboflavin are not lost during storage of these commodities prior to processing. Although present in small quantities, these vitamins are very stable in potatoes during storage (Table 6).[8]

Carotene

Fresh fruits and vegetables show marked variation in carotene content according to variety. There is little evidence of any loss of carotene during ordinary storage and handling for market unless adverse conditions prevail. In some commodities, such as the sweet potato, there is an actual synthesis and total increase in carotene during storage.[9] However, leafy vegetables subjected to wilting conditions may lose one-half to two-thirds of their carotene when held at or near room temperature for 4 days.[10]

Carbohydrates

The change in sugar content in freshly harvested sweet corn is often referred to as a classic example of the effect of temperature on changes in composition of a freshly harvested commodity. In recent surveys reported from Pullman, Washington, on the commercial handling of sweet corn for processing where transit and holding times of the loads ranged from 3½ to 20 hours and the average temperature of the loads ranged from 50° F. to 90° F., the loss in sugar ranged from 1 to 27 percent.[11]

The hydrolysis of a portion of the starch to sugar in potato tubers stored at temperatures below a certain level, usually

Table 5.—Loss of Thiamin and Riboflavin in Shelled Lima Beans at Room Temperature. (After Eheart et al[5])

| | Average loss (percent) | |
	Thiamin	Riboflavin
2 days	9	14
4 days	10	7

Table 6.—Effect of Storage on Nutritive Value of White Rose Potatoes. (After Yamaguchi et al[8])

	Vitamins				Minerals		
	C	B₁	B₂	Niacin	Ca	Fe	P
	mg per 100 grams edible portion						
At harvest	36	0.12	0.025	0.66	14	2.0	53
41°F storage							
6 weeks	19	0.13	0.032	0.70	13	1.8	49
12	16	0.12	0.034	0.62	13	1.7	46
18	11	0.12	0.027	0.60	12	1.9	49
24	10	0.11	0.031	0.56	11	1.9	54
30	10	0.13	0.027	0.56	12	2.6	54
50°F storage							
18 weeks	12	0.13	0.027	0.60	14	1.7	52
24	10	0.11	0.029	0.58	14	2.0	56
30	8	0.11	0.026	0.60	14	---	55

about 50° F. is another example of the effect of temperature on composition.

Many other changes such as the synthesis and hydrolysis of proteins and changes in pectic constituents during storage could also be described but it is doubtful that these changes have any real nutritional significance. Some of the changes may have marked effects on the acceptance quality. The loss of sugar in sweet corn and its increase in potatoes are both detrimental to quality while the increase in sugar in sweet potatoes and parsnips is beneficial.

Nitrates and Nitrites

Another change that may occur in some commodities during storage is a conversion of nitrates to nitrites. Recent studies emphasize the seriousness of these changes. It has been shown that heavy application of nitrogen fertilizer during the production of spinach will increase the nitrate concentration and that 10 to 15 percent of the nitrates may be converted into nitrites after spinach has been held at room temperature for 4 days

(Table 7).[12] Other studies with beets, spinach (Table 8)[13], and potatoes[14] have shown that the levels of nitrites remain low at different fertilization levels. The conversion of nitrates to nitrites was minimal or nonexistent in beets and spinach when they were stored at 35° F. or when potatoes were held at a desirable storage temperature. However, potatoes in atmospheres containing low oxygen concentrations showed a slight increase in nitrites.

NEW STORAGE TECHNIQUES

So we may ask, what new storage techniques are being used today that have an influence on the composition or nutritional quality of fruits and vegetables? The use of modified or controlled atmospheres for prolonging the storage of apples has been a commercial practice for a number of years and has been expanding for apples and for other commodities in recent years. A study in France indicates that apples stored in 3 percent oxygen and 97 percent nitrogen at 15° C. retained over 60 percent of the original ascorbic acid after 3 months of storage, whereas those stored in air retained less than 20 percent (Table 9).[15] When similar tests were conducted at a more normal storage temperature of 4° C. no differences in ascorbic acid retention were noticeable between the lots stored in controlled atmosphere and in air. Spinach stored at 45° F. in an atmosphere containing 4 percent oxygen and 9.2 percent carbon dioxide for 8 or 9 days retained 72 percent of its ascorbic acid whereas spinach stored in air at the same temperature retained

Table 7.—Nitrate and Nitrite Content of Spinach at Harvest and After 1 and 4 Days Transportation or Storage. (After Schuphan et al[12])

Nitrogen Fertilizer Levels	mg N per 100gm fresh weight					
	At Harvest		After 1 Day		After 4 Days	
	NO_3	NO_2	NO_3	NO_2	NO_3	NO_2
N_1	12.9	0.03	14.2	0.07	13.7	0.33
N_2	103.	0.26	121.	3.75	132.	18.4
N_3	236.	0.20	243.	2.43	210.	29.20

Table 8.—Nitrate and Nitrite-N Content of Beets and Spinach Grown at Different Levels of Nitrogen Fertilizer. (After Lee et al[13])

Fertilizer lb/acre	Nitrate-N mg/100gm	Nitrite-N ppm
Beets		
0	40	0.9
100	110	1.5
200	310	1.2
400	350	2.0
Spinach		
0	248	0.5
400	690	0.6

Table 9.—Effect of Controlled Atmosphere Storage at $15°C$ on Ascorbic Acid Content of Apples. (After Delaporte[15])

Days in storage	mg/100gm fresh weight Air	$97\% N_2 + 3\% O_2$
8 to 10	18.1	24.1
30 to 35	8.9	18.4
58 to 66	5.5	15.9
72 to 85	3.3	14.9

approximately 45 percent.[16] When spinach was stored at a more desirable temperature of 34° F. under controlled atmosphere and in air, very slight differences in ascorbic acid retention were noticed (Table 10). Atmospheres modified by the addition or the controlled accumulation of carbon dioxide have been found to aid in retaining the sugar content of sweet corn and the tenderness of asparagus.

Recent experiments at the University of Maryland have included the use of toxic gases, carbon monoxide and ethylene oxide, for preserving the fresh quality of mushrooms and sliced peaches. Compositional analysis indicated that the thiamin con-

Table 10.—Changes in Ascorbic Acid Content of Spinach Stored in Air and in Controlled Atmospheres (CA) at Two Temperatures. (After Burgheimer et al[16])

Storage time days	34°F		45°F	
	Air	CA*	Air	CA*
	mg per gm dry matter			
0	6.35	6.35	7.15	7.35
1	5.88	5.15	6.65	6.57
3	4.82	4.47	5.21	6.63
5	4.08	4.39	4.42	6.44
7	3.90	3.45	3.19	5.33

*9.2 percent carbon dioxide and 4 percent oxygen

tent of mushrooms stored in carbon monoxide atmospheres for 1 month increased 100 percent.[17] No other major changes were noted in such components as total protein, relative protein value and sugars.

The use of hypobaric or low pressure storage for fruits and vegetables is being investigated in several laboratories. Its effectiveness in prolonging the storage life of some commodities is evident but the influence on composition is not known at this time.

Much research has been conducted recently on ethylene and its role in plant physiology. Ethylene is used commercially on a number of fruits and vegetables. It has been used for years to hasten and improve the uniformity of ripening of bananas. It is also used to stimulate the ripening of tomatoes and honeydew melons and to degreen citrus fruits. It is generally accepted that ethylene functions as a plant hormone and that one of its roles is a triggering of the ripening process. There is no substantial evidence to indicate that fruits stimulated to ripen because of added ethylene differ significantly in nutritional value from those ripened without added ethylene.[18]

In summarizing, we can say that the losses in nutritive value of fresh fruits and vegetables during storage, transportation and marketing are not likely to be great if the following are observed:

- Care is taken to keep physical damage at a minimun during handling.
- The time between harvesting and receipt by the consumer is not prolonged.
- The temperature and humidity conditions are kept near optimum for the commodity.

References

1. Ezell, BD and Wilcox, MS: Loss of vitamin C in fresh vegetables as related to wilting and temperature, *J Agr Food Chem* **7**:507-509, 1959.
2. Lieberman, M; Craft, CC; and Wilcox, MS: Effect of chilling on the chlorogenic acid and ascorbic acid content of Porto Rico sweet potatoes, *Proc Amer Soc Hort Sci* **74**:642-648, 1959.
3. Murphy, E: Storage conditions which affect the vitamin C content of Maine-grown potatoes, *Amer Potato J* **23**:197-217, 1946.
4. Werner, HO and Leverton, RM: The ascorbic acid content of Nebraska-grown potatoes as influenced by variety, environment, maturity, and storage, *Amer Potato J* **23**:265-267, 1946.
5. Eheart, JF et al: Vitamin studies on lima beans, bulletin 5, Southern Coop. Series, 1946.
6. Hulme, AC: *The Biochemistry of Fruits and Their Products*, New York: Academic Press, 1971, vol 2, chap 3.
7. Fager, EEC et al: Folic acid in vegetables and certain other plant materials, *Food Res* **14**:1-8, 1949.
8. Yamaguchi, M; Perdue, JW; MacGillivray, J: Nutrient composition of White Rose potatoes during growth and after storage, *Amer Potato J* **37**:73-76, 1960.
9. Ezell, BD and Wilcox, MS: Effect of variety and storage on carotene and total carotenoid pigments in sweet potatoes, *Food Res* **13**:203-212, 1948.
10. Ezell, BD and Wilcox, MS: Loss of carotene in fresh vegetables as related to wilting and temperature, *J Agr Food Chem* **10**:124-126, 1962.
11. Smittle, DA; Thornton, RE; and Dean BB: Sweet corn quality deterioration, circular 555, Washington Agriculture Experiment Station, 1972.
12. Schuphan, W and Schlottman, H: N-Uberdüngen als Ursache hoher Nitrat- und Nitritgehalte des Spinats (Spinacia oleracea L.) in ihrer Beziehung zur Sauglings-Methamoglobinamie, *Z Lebensmittel Untersuchung und-Forschung* **128**, 71-75, 1965.
13. Lee, CY et al: Nitrate and nitrite nitrogen in fresh, stored and processed table beets and spinach from different levels of field nitrogen fertilization, *J Sci Food Agr* **22**:90-92, 1971.

14. Hata, A and Kuniyasu, O: Changes of nitrate and nitrite contents during storage and cooking of potato tubers, *J Jap Soc Food & Nutr* 24:345-349, 1971.

15. Delaporte, N: Influence de la teneur en oxygene des atmosphères sur le taux d'acide ascorbique des pommes au cours de leur conversation, *Lebensm-Wiss u Technol* 4:106-112, 1971.

16. Burgheimer, F; McGill, SN; Nelson, AI; Steinberg, MP: Chemical changes in spinach stored in air and controlled atmosphere, *Food Technol* 21:1273-1275, 1967.

17. Besser, T and Kramer, A: Changes in quality and nutritional composition of foods preserved by gas exchange, *J Food Sci* 37:820-823, 1972.

18. Harris, RS and Von Loesecke, H: *Nutritional Evaluation of Food Processing*, Westport, Conn: AVI Pub Co, 1960, chap 2A, pp 86-89.

Additional References

Barker, J and Mapson, LW: The ascorbic acid content of potato tubers. II. The influence of the temperature of storage, *New Phytol* 49:283-303, 1950.

Boggers, TS, Jr; Marion, JE; Dempsey, AH: Lipid and other compositional changes in nine varieties of sweet potatoes during storage, *J Food Sci* 35:306-309, 1970.

Craft, CC and Heinze, PH: Physiological studies of mature-green tomatoes in storage, *Proc Amer Soc Hort Sci* 64:343-350, 1954.

Ezell, BD and Wilcox, MS: Influence of storage on carotene, total carotenoids and ascorbic acid content of sweet potatoes, *Plant Physiol* 27:81-94, 1952.

Ingalls, R et al: The nutritive value of canned foods. III. Changes in riboflavin content of vegetables during storage prior to canning, *Food Tech* 4:258-263, 1950.

Ingalls, R et al: The nutritive value of canned foods. IV. Changes in thiamine content of vegetables during storage prior to canning, *Food Tech* 4:264-268, 1950.

Lougheed, EC and Dewey, DH: Factors affecting the tenderizing effect of modified atmosphere on asparagus spears during storage, *Proc Amer Soc Hort Sci* 89:336-345, 1966.

Phillips, WEJ: Changes in the nitrate and nitrite contents of fresh and processed spinach during storage, *J Agr Food Chem* 16:88-91, 1968.

Shiota, Y and Kurogi, M: Post-harvest changes of oxidases, ascorbic acid, and total phenols in cabbage heads, *J Food Sci and Tech* (Jap) 16: 135-139, 1969.

Tolle, WE: Hypobaric storage of mature-green tomatoes, Marketing Research Report No 842, U.S. Department of Agriculture, 1969.

Ulrich, R and Delaporte, N: L'acide ascorbique dans les fruits conserves par le froid, dans l'air et en atmosphère controlée, *Ann Nutrition Alim* 24:B287-B325, 1970.

The Prospects for Genetic Engineering to Improve Nutritional Values

W. H. Gabelman, Prof.

"Genetic engineering" is a term somewhat synonymous with "breeding". Both imply the development of new biological materials that fit a need; however, the former term suggests orderly, systematic and predictable change. "Breeding" is a much less precise term and implies that new cultivars can be developed but without necessarily describing the scientific precision and methodology.

There are many factors which determine whether or not a newly developed cultivar, irrespective of its nutritional value, will be used by the public in our freely competitive society. If it will not be used, the potential for nutritional importance is nil.

The cultivars must be grown by *someone*; the farmer (or grower) must make a profit or reduce his risks. He is concerned with yield, uniformity, adaptation to mechanization, pest resistance, storage quality, and appearance (for sale purposes). *Someone* must handle the product at the farm, in transit, and at wholesale and retail outlets. These people are concerned with uniformity, adaptation to mechanical handling, appearance and freedom from defect. *Someone* must buy the product at the retail level. A highly diverse group of interests buy for reasons of cost, impulse, appearance, personal food habits, apparent "freshness" of product, and food value—probably in that order. *Someone* must consume the product. We tend to "taste" the product with our eyes and noses as well as our tongues. If need be, we can add sugar, salt and butter—we can always cover it with a "tasty" sauce.

These are the people who determine the cultivars that will be grown. It is a haphazard, non-ordered system of decision-making. In nearly all instances, the fruit or vegetable which can be merchandised *attractively* at *the lowest cost per unit of sale* will survive. All other factors are academic. If you find this difficult to believe, talk to the farmer who must sell his product for a living. So let us be realistic as we assess our *prospects.*

In the past, nutritional quality has not been an important factor in decision-making on the farm and in the market place. Unfortunate, but true. We eat oranges for vitamin C but we do not buy one cultivar of oranges because it has more or less vitamin C than any other. Until we change our marketing system, nutritional quality will remain the hope of idealists.

A genetic engineer, much like an architect, designs new plants that are functional. He must recognize all limiting factors in production and marketing because only one weakness will cause the elimination of the new cultivar. These factors have highest priority. If time, energy and resources permit, other highly desirable, though less critical, factors can be considered. Nutritional quality fits into this latter category.

Costly research procedures tend to be eliminated in favor of less expensive programs. Nutritional evaluation must be listed as costly. Genetic engineers working for industrial concerns must show a profit either from seed sales or processed product, depending on their source of support. Nutritional quality has never demonstrated any great leverage here.

Genetic engineers at public institutions, who should be and often are concerned with nutritional quality, do not have the agricultural research resources needed to do a reasonably effective job. Both the National Institutes of Health and National Science Foundation have felt that genetic engineering was a responsibility of the U.S. Department of Agriculture and state agricultural experiment stations. We are caught in the middle of public desire and public responsibility.

So much for the negative viewpoint. The composition of the plant is our primary concern here, so let us review our state of knowledge in three areas: carbohydrate and ascorbic acid synthesis, protein synthesis, and, then, in a bit more detail, carotene synthesis. Please keep in mind that genetic engineers require variation that is under gene control. These variations are the building blocks of new cultivars. Keep in mind also that genetic engineers need to work with individual plants within large segregating populations and that effective, precise methods of discriminating superior and inferior individuals, in a way that can be applied to large populations, is a prerequisite to efficiency.

CARBOHYDRATE AND ASCORBIC ACID SYNTHESIS

One primary reason for eating is to provide energy for the life processes. Certain of our fruits and vegetables, e.g., apples, potatoes, sweet potatoes, and corn, are considered to be important dietary sources of energy.

We should be impressed with our systematic gain in yield of most crops. Part of this gain has a genetic base and part relates to our ability to manipulate the plant culturally or change its environment. Yield is usually a terminal growth measurement and, therefore, is quite an inefficient measurement in breeding systems. We now can measure heritable differences in (1) the efficiency of CO_2 fixation, (2) different (C_3 and C_4) systems of CO_2 fixation, and (3) net photosynthesis, using leaf discs or young plants. We hope these new techniques will relate to significant aspects of quality and yield because they can be assayed early in the life cycle. Everything must start with photosynthesis.

Many other facets of plant growth are directly related to the fate of the fixed carbon. Sweetness is associated with the amount of reducing sugar, not total sugars. The heritability of these differences in sugar has been well documented in carrots and is generally appreciated in all horticultural crops. For many years, we have known that the difference between starchy field corn and sweet corn is a single gene *su* which slows the rate of conversion of sugars to starch. There are a number of different *su* genes in maize. When *su* is combined with the gene sh_2, a super sweet endosperm results. As a result, new cultivars of predetermined types of carbohydrates are rapidly developed.

Ascorbic acid is a rather simple carbohydrate. It is extremely ubiquitous in plants; yet we only think of certain plants such as citrus and tomatoes as significant dietary sources.

The direct formation of ascorbic acid from hexoses in plants was reported over 35 years ago. In general, we know the biosynthetic pathway but do not know why citrus and tomatoes have such high levels. The total amount of ascorbic acid in tomatoes and potatoes (10 to 30 mg/100 gm fresh weight) is heritable and development of strains with significantly more or less ascorbic acid is practical.

PROTEIN SYNTHESIS IN SEEDS

Legumes and maize represent significant sources of protein in human diets. We have long known that total protein of maize could be increased greatly by breeding but that the cow could not recognize the difference; increasing the lysine-deficient zein was of no great value. Rats grown on zein died unless supplemented with tryptophan and lysine. The glutelin fraction of maize protein is over three times as rich in lysine as zein. Mutants of maize which alter the ratio of prolamines to glutelins greatly alter the nutritional value. Higher animals can detect this difference.

In legumes, the globulins, which are the major protein fraction, are deficient in methionine and cysteine, whereas the albumin fraction is much richer in both sulfur-containing amino acids. Shifting the ratio of globulin to albumin is important. Kelly[1] has demonstrated a wide range of microbiologically available methionine among 3600 different beans and has also demonstrated the heritability of these differences. He also stated that "if methionine levels were doubled their (bean) amino acid balance would approach the highly desirable egg protein pattern for amino acids. Such increases are realistically attainable".

CAROTENE SYNTHESIS

Carotenes, like ascorbic acid, are ubiquitous in plants. Their role in plant metabolism is not obvious.

I would like to detail some of our results from studying carotenes in carrots since these results tend to embody much of my message and does have broad applicability to both fruits and vegetables.

If the historical records embodied in the works of the old painters are accurate, orange carrots are a product of the last 500 years. Until 1500 A.D., only white carrots were painted; then yellow carrots appeared. Orange carrots were first seen about 1700 A.D. This tells us something about the genetic system. Since organisms are more apt to lose a genetic function via mutation than gain one, we can assume that orange must be the loss of function and, therefore, probably recessive.

When we started our carrot improvement program in 1950, we knew a few things about color. We knew that carotenes were the primary pigments in carrot roots. We knew that core and cortex seldom were the same color and varied greatly. We knew that white root phenotypes were dominant to colored phenotypes. We also knew the literature to be full of suggestions that the environment of the plant greatly affected carrot color. The Wisconsin carrot industry needed better colored carrots in order to remain competitive. No one asked for carrots with more provitamin A.

We went to the commercial fields and spent a large amount of time selecting roots of many different colors. Self-pollinated progeny of these roots were evaluated. We needed to gain some insight into the cause of color variation. What was genetic and what was non-genetic? We found the interior color of the roots varied around the value of the parent root, indicating that the primary cause of variation was genetic, not environmental. Uniformly dark orange roots produced progeny which did not segregate for color. This suggested that dark orange phenotypes were the expression of a homozygous recessive set of genes.

Among the dark orange interiors there was no evidence of internal greening. On the basis of the above listed data (results from very simple experiments), we were able to visualize the genetic engineering procedures needed and could justify the program.

Briefly, our program for color included two key procedures: (1) inbreeding and selecting for the dark orange types and (2) hybridizing the dark orange inbreds to produce a *vigorous* and *uniform* F_1 hybrid of desirable color. The initial objectives have now been reached in both cases.

We recognized quickly that an inbreeding program gave us a

unique opportunity, as public servants, to become much more knowledgeable about the biosynthesis of the various carotenes. After all, not all carotenes convert equally well to vitamin A. So we added a dimension beyond the need for color *per se*.

Commercial cultivars of carrots have about 100 to 120 μg/gm fresh weight of carotene. About 60% is beta carotene, 20% alpha carotene and about 20% is a mixture of colorless polyenes and colored carotenes, particularly lycopene, zeta carotene and gamma carotene.

We found that white, which had previously been reported as a single gene dominant, reduced total carotenoids from 120 μg/gm to <1 μg/gm. There was no provitamin A potential at all. We also found that the intensity of visual orange coloration was associated with total carotenoid synthesis. Types ranging from 1 μg/gm to 400 μg/gm total carotenoids were isolated. Beta and alpha carotenes were the primary pigments. However, the ratio of beta to alpha changes as total carotenoids increase.

Surprisingly, we found three independently inherited genes which limited the amount of total carotenoids synthesized as follows:

1. yy rr ss   120 μg/gm
2. yy rr s   60 μg/gm
3. yy r   30 μg/gm
4. y   0-1 μg/gm

Strong epistasis was evident with Y epistatic to R; R epistatic to S. The range in total carotenoids synthesized for each genotype was surprisingly narrow. We have no evidence that gene R or S alters the ratio of pigments, rather all of our evidence suggests an effect on a common substrate that allows the normal controls, in the production of specific pigments, to operate thereafter.

At this stage, we had identified three genes and had established a potential range in pigment production that extended well beyond the levels characteristic of commercial cultivars.

We were also concerned with differences in color of core and cortex. There were progeny which segregated differentially for this color. Some progeny were uniformly colored throughout the root. In a rather exhaustive study, we were able to identify five genes which differentiated color of the White Belgian carrot and an orange Wisconsin inbred, W93.

Briefly the five genes gave phenotypes as follows:

Y = white

yy = colored

yy Y_1 or Y_2 = colorless or lightly colored cores

yy y_1y_1 y_2y_2 = uniform core and cortex color

yy y_1y_1 y_2y_2 IoIo OO = uniformly dark orange

yy y_1y_1 y_2y_2 ioio oo = must be white!

Thus, we had isolated two new genes which gave rise to differential core color (Y_1 and Y_2) with the dominant alleles repressing pigment development. The genes Y, Io and O may have been alleles of the three reported for total carotenoid synthesis earlier (Y, R and S). Thus far, all of the genes isolated altered pigment formation quantitatively.

In our measurement of pigments, we found types that seemed to have *ratios* of beta to alpha *differing* from the 3 : 1 ratio characteristic of most varieties. In 1957, we systematically analyzed each of five roots per line of 200 different breeding lines of carrots for total carotenoids, beta carotene and alpha carotene. The mean of the ratios indicated that the ratio of beta to alpha decreased as total carotenoids increased. When examining the specific behavior of each of the 200 lines, we found that:

(1) the ratio of beta to alpha was stabilized with relatively little inbreeding, indicating simple inheritance for this trait; and

(2) the synthesis of alpha carotene never exeeded the synthesis of beta carotene although lines with 1 : 1 ratio were not infrequent and could be found at all levels of total carotenoid synthesis from 40 to 200 μg/gm.

We found some lines that produced 85-90 μg/gm fresh weight of alpha carotene, which is very high, but they always produced an equivalent amount of beta carotene. The biological significance of this 1 : 1 ratio has not been explained. Wide beta to alpha ratios could also be found. For example, at low total carotenoid levels (30-50 μg/gm), ratios as high as 35 : 1 could be found; at high total carotenoid levels, ratios of 8 : 1 were maximum. The maximum ratio seemed to be correlated with the capacity for total carotenoid synthesis. For reasons of money, this problem has been relegated for study at a later date.

Although the primary colored carotenes in carrots are beta and alpha carotene, we do have significant amounts of gamma carotene, zeta carotene and lycopene in certain lines. The Red Japanese variety, Kintoki, has nearly the same pigment distribution as we find in tomato, namely, *very high lycopene*, high beta carotene and *no alpha carotene*. The Wisconsin inbred, W93, has alpha and beta carotene but no lycopene. Here then was an opportunity to study the dependence or independence of these two pigments using our carrot material as the test organism.

Briefly, we mated: W93 alpha$^+$beta$^+$lycopene$^-$
            with Kintoki alpha$^-$beta$^+$lycopene$^+$
The F_1 was alpha$^\pm$, beta$^+$, lycopene$^\pm$, and was orange (visually).

The F_2 segregated 13 orange to 3 red and this was substantiated by appropriate test crosses as a digenic control as follows:

9 alpha$^\pm$beta$^+$lycopene$^\pm$orange
3 alpha$^+$beta$^+$lycopene$^-$orange
3 alpha-beta$^+$lycopene$^+$red
1 alpha$^-$beta$^+$lycopene$^-$orange

Three things are significant: (1) Alpha carotene and lycopene are independently inherited. (2) Alpha carotene masks the red of lycopene to make carrots appear orange rather than red. (3) Isolates in which the primary colored pigment is beta carotene can be recovered. It does not appear from our results that the elimination of alpha carotene and lycopene synthesis enhanced the total beta carotene produced.

In summary, we now know considerably more about the genetic controls of carotene synthesis than we did before we started 20 years ago. Because of this, we now can carry out a program in genetic engineering that is logical, efficient and productive. In our studies, we have isolated and identified 14 genes controlling color. At least six of the genes have been presented here as follows:

Y _ = white

yy R _ = 30 μg/gm total carotenoids

yy rr s _ = 60 μg/gm total carotenoids

yy Y_1— or Y_2— = differential core/cortex color

yy y_1y_1 y_2y_2 = uniform core/cortex color

yy y_1y_1 y_2y_2 IoIo 00 = uniform intense orange pigment

yy y_1y_1 y_2y_2 IoIo 00 R R = uniform intense orange pig-

yy y_1y_1 y_2y_2 IoIo oo R _ = uniform intense red - no alpha

yy y_1y_1 y_2y_2 IoIo 00 rr = uniform intense orange - no lycopene

yy y_1y_1 y_2y_2 IoIo oo rr = uniform intense orange - only beta

yy y_1y_1 y_2y_2 ioio oo rr = (not identified) but must be white!

Our next project, which can be very significant in breeding, is the isolation and identification of the total recessive genotype. This material should act to mirror, in F_1 hybrids, any introduced dominant genes for pigmentation. Immediate identification of genotypes is always near the top of priorities to genetic engineers.

I hope our results on carrots will help justify support of similar, appropriate genetic studies in higher plants. We need information on the biochemical genetics relating to synthesis of nutritionally important compounds. We need assay procedures which are quick, inexpensive and precise. We need the opportunity to do the genetic engineering on important compounds that we know man needs and we need the opportunity to work on the nutritional factors that bioassays tell us are present but which have not been documented chemically.

In short, we need laboratories of excellence with the complementary disciplines in nutrition, genetics and plant breeding committed to the nutritional *needs* of mankind. There is really nothing so very complex or difficult that substantial progress cannot be made in this field, but if other priorities in the market place prevent the use of materials of greater nutritional value, it will be difficult to excite the biologists of the future in this area.

References

1. Kelly, JF: *J Amer Soc Hort Sci* 96:561-563, 1971.

Genetic Engineering to Remove Undesirable Compounds and Unattractive Characteristics*

August E. Kehr, Ph.D.

Traditionally the greatest emphasis in plant breeding has been placed on increasing our total food supply. As a result of the success of this plant breeding research, food shortages are considered less of a threat throughout the world today than in the 1960's. Consequently, the average well-fed American today would probably give top priority to the problems of consumer acceptability and food safety.

NATURALLY OCCURRING TOXICANTS IN PLANTS

The recent public health concern has arisen from the growing realization that many foods contain naturally occurring toxic or deleterious properties. As new germ plasm is used to breed vegetables and other food crops to increase resistance to insects and diseases, there is real danger of increasing levels of these undesirable substances. For example, a potato variety released because of its superior processing qualities had an unusually high content of total glycoalkaloids. Likewise, some potato cultivars are insect resistant because they contain a high content of a cholinesterase inhibitor.

Although there is great potential public health significance to the presence of substances such as alkaloids, antimetabolites, oxalates, and allergens in foods, much too little research has

---

* Official information material, U.S. Government; this chapter may be reproduced in part or in full.

been devoted to the genetic or physiological factors that determine their synthesis and ultimate level in plants. Because research in these areas is relatively new, the total genetic knowledge developed and recorded in the literature is frustratingly sparse. Few reports are to be found that define an association between genotype and undesirable or unattractive characters. We might note that most of the papers on this subject have been published since 1970.

NITRATES

The hazard of excessive nitrates in foods has been generally recognized, especially for infants.[1]

Nitrates can be present in significant quantities in some vegetables under circumstances such as nitrogen-rich soils and reduced light and water. Under these conditions, lettuce, turnips, greens, collards, dryland cress, beets, radishes, celery, endive, kale, and parsley can have relatively high concentration of nitrates.[2] On the other hand, the concentration in tomatoes, potatoes, carrots, beans, peppers, sweet corn, and peas is quite low. Viets and Hageman,[3] in making a study of factors affecting the accumulation of nitrates, found in the literature little or no specific data concerned with differences in nitrate absorption as a function of genotype. However, genetic control of nitrate accumulation in vegetables has recently been demonstrated. Small but consistent differences were found in nitrate content of the spinach cultivars, Winter Bloomsdale, Northland, and Tuftguard, fertilized at four rates of nitrogen.[4]

Winter Bloomsdale, a savoy-leaved cultivar, accumulated nearly twice the nitrate found in Northland and Tuftguard, the two smooth-leaved cultivars, indicating that the uptake of nitrogen may be accidentally associated with the genes for savoy-leaf. Savoy-leaf is partly dominant over smooth leaves.

The association of nitrate N uptake and leaf character was further verified by Maynard and Barker[5]. (See table on following page)

However, Cantliffe[6] could not detect consistent differences in nitrate between savoy- and smooth-leaved cultivars, although he did find that under conditions of nitrate fertilization the cultivar, Virginia Savoy, contained more nitrate than the

Association of Nitrate N Concentration With
Leaf Type in Spinach

Leaf Type	No. of Cultivars	Nitrate N(% dry wt.)	
		Petioles	Blades
Savoy	6	1.14 b	.30 b
Semi-savoy	6	.90 b	.18 ab
Smooth	6	.42 a	.08 a

smooth-leaved cultivar, Northland. He concluded that the variation in nitrate content was not caused by differences in nitrogen uptake because both cultivars contained a similar concentration of total nitrogen. However, the differences in nitrate content could have been caused by a genetic difference in nitrate reductase activity between the two cultivars, with no real association between leaf type and nitrate reductase activity.

In a cooperative study involving seven universities, cultivar characteristics were a major factor controlling nitrate accumulation in three vegetables.[7]

Nitrate content in plants varies inversely with the level of nitrate reductase activity in the same tissue.[8] The level of this reductase activity is highly heritable.[9] These findings demonstrate the feasibility of breeding many plants that will have relatively higher levels of nitrate reductase and hence lower levels of nitrates in the tissues. In summary, genetic control of nitrate uptake or reduction, or both, has been demonstrated in vegetables, including the feasibility of developing cultivars low in nitrate content.

NEURO-TOXINS AND LATHYROGENS

Lathyrism in man is characterized by muscular weakness and paralysis of the lower legs, disabling the victim and making it impossible for him to walk. This disease is associated with the consumption of seeds of *Lathyrus* species and is found in some areas of India and the Mediterranean area. In India cultivation of this crop is banned, but it still persists as a weed in grain fields because of its high drought resistance and consequently is eaten when the main crop fails. The problem of lathyrogens in foods has been carefully reviewed by Liener.[10]

According to Swaminathan,[11] lathyrism is caused by the neurotoxin β-N-oxalyl-$\alpha\beta$-diaminopropionic acid (BOAA). He reported that natural populations of *Lathryus sativus* contained from 0.5 to 2.51 percent of the toxic principle.

Swaminathan and his coworkers treated seeds of four commercial cultivars of *L. sativus* with X-rays and nitrogen mustard and selected individuals in the M_2 generation for low content of BOAA. The mutants were true breeding for low content of BOAA. These individuals were then subjected to a second cycle of treatment to isolate mutants even lower in content of BOAA. On the basis of data collected, the production of the neurotoxin was controlled by only a few genes. Apparently lines of *Lathyrus* could be developed that are essentially devoid of the toxic principle.

GLYCOALKALOIDS

Glycoalkaloids are a naturally occurring constituent of all potatoes as well as other vegetables such as eggplant, peppers, and tomatoes. Alkaloids impart a bitter flavor to potato tissues and may be a factor in protecting plants from insects[12] and fungi.[13] Some instances have been reported, primarily in Europe, where excessive amounts in potatoes have caused intestinal disorders and even death to humans and livestock.[14]

In the normal potato tuber, most of the glycoalkaloid content is found near the skin and is consequently removed upon peeling, as shown in the following data:

TGA Concentration (mg/100 gm fresh wt.)

Potato Variety	Peel	Flesh	Peal and Flesh
Katahdin	81.0	2.3	10.1
Kennebec	76.0	1.5	9.7
Russet Burbank	69.0	1.2	8.0

The problem of glycoalkaloids in potatoes came to the fore-

front recently with the discovery of unusually high content in
the potato breeding line B5141-6, formerly, Lenape. Lenape
was released in 1967 by the U.S. Department of Agriculture and
the Pennsylvania Agricultural Experiment Station because of its
high dry matter and adaptability for processing. Two years later
scientists discovered that this variety consistently had a total
glycoalkaloid content about twice as high as that of any other
variety, as follows:

Variety	TGA mg/100 gm	
	1968	1969
Katahdin	10.1	12.1
Lenape (B5141-6)	18.1	25.4

(Unpublished data from Dr. S. L. Sinden—USDA)

As soon as it was realized that this variety had higher than
normal glycoalkaloid content, the USDA and the Pennsylvania
AES took immediate steps to remove this variety from the
commercial trade.[15]

Primarily as a result of the Lenape episode, research on
glycoalkaloids in potatoes was expanded and new research was
initiated in both the United States and Canada, particularly on
the nature of inheritance of glycoalkaloid content in potatoes.
Sanford and Sinden[16] found significant differences in tuber
glycoalkaloid (TGA) content among parents and among family
means in a 2-year study of 10 tetraploid crosses. TGA contents
of the parents ranged from 3.6 to 36 mg/100 gm, with an
average content of 10 mg/100 gm. The average content in
commercial varieties is about 8 mg/100 gm.

Offspring variations within families were generally continuous, indicating polygenic inheritance. Sanford and Sinden concluded that TGA content was highly heritable and hence potato
breeders could maintain low TGA contents in their breeding
lines or even breed for a lower content of TGA.

Orgell and Hibbs[17,18] studied *in vitro* cholinesterase inhibition by potato tissue extracts in many potato varieties and
species. Extremely high human plasma cholinesterase inhibition
was found in many species of wild potatoes, as well as a lower

but wide range of readings among commercial cultivars. They
believe that plasma cholinesterase inhibition probably reflected,
at least in part, the presence of steroid alkaloids such as solan-
ine.

Potatoes exposed to light turn green and become bitter as a
result of increased levels of chlorophyll and glycoalkaloids.
However, recently potato clones have been identified that,
when exposed to light, do not turn green at intensities found in
homes, stores, and storages. Incompletely dominant multiple
genes are involved in the inheritance of tuber greening.[19] Studies have also been made to define the glycoalkaloid content in
these greening resistant clones[20] because light induces both
glycoalkaloid and chlorophyll synthesis. Tubers of these clones
synthesized less total glycoalkaloids and developed less chlorophyll than did tubers of clones that lacked resistance to greening. In similar manner genetic variability for nongreening occurs
in other vegetables, including carrots. The interrelationships of
glycoalkaloid formation, cholinesterase inhibition, and nongreening require additional clarification before their significance
can be ascertained, but we now have sufficient information on
their genetic control to breed and select desirable types in this
regard.

OXALATES

Soluble oxalates when fed to 50-day-old rats completely
immobilized body calcium.[21] However, after a careful review of
the literature, Fassett[22] questioned whether the reports on
hazards to humans from the ingestion of oxalates in rhubarb
leaves or in oxalate-containing vegetables seem warrnated. On
the other hand, Kingsbury[14] categorically states that ingestion
of relatively small amounts of the rhubarb leaf blade is potentially lethal.

A study of 20 cultivars of spinach revealed highly significant
differences in oxalic acid content, ranging from 8.66 percent to
10.52 percent on the dry-weight basis.[23] These results suggest
that spinach can be selected and bred for low oxalate content.
However, the low oxalate lines of spinach are reported flat in
taste. Because of the uncertainty on the hazards of oxalates and

also the flat taste factor, low oxalate spinach perhaps may not be a desirable breeding objective. In any case we know of no serious attempt to develop low-oxalate cultivars.

FLATULENCE

Dry beans, lima beans, and related plants are often called the poor man's meat. They are a cheap source of protein for much of the underprivileged populations of the world, expecially Latin America and Southeast Asia. However, many dry seeds of the legume family produce discomforting flatus when eaten by humans. Dry navy, kidney, and pinto beans produce especially high levels of flatulence.[24] The ratio of flatus from a bland test meal of 100 gm (dry weight) measured for a 3-hour period, (from 4 to 7 hours after ingestion) as compared with flatus from different legumes is as follows:

Flatus From Different Legumes

	Ratio
Phaseolus vulgaris L.	
California small white	11.1
Pinto	10.6
Kidney	11.4
Phaseolus lunatus L.	
Lima, Ventura	4.6
Lima, Fordhood	1.3
Phaseolus aureus Roxb.	
Mung	5.5
Glycine max. (L.) Merr.	
Soya, Lee or Yellow	3.8
Arachis hypogaea L.	
Peanut	1.2
Pisum sativum L.	
Pea, dry	5.3
Pea, green	2.6
Bland test meal	1.0

Murphy also reported that there were significant differences in cultivars of dry lima beans in the flatulence factor, as follows:

Variety	Average Flatus in 3 Humans
	(Total vol. in cc for 3-hr. peak period)
Fordhook - Green mature	18
Fordhook - Dry	39
Ventura - Dry	227
Bland Formula	23

The lima bean cultivar, Fordhook (*Phaseolus lunatus* L.), was developed for use as a frozen product and is seldom eaten in the dry form, whereas the cultivar, Ventura, is primarily grown for the dry product. Therefore, the flatulence factor has been either lost in the development of the Fordhook cultivar or inadvertently added in the development of the Ventura.

An heirloom dry bean variety (*Phaseolus vulgaris* L.) has been kept in the Pike Family in New England for many years under the name "Jacobs Cattle" bean. Murphy[24] reported that this bean raised the formation of intestinal gas about four times that of the bland control as compared with the expected ten-fold response to the usual dry *Phaseolus vulgaris* group.

LOW-DIGESTIBLE STARCH

The quality of fresh sweet corn depends to a large extent on the sugar and polysaccharide content of the endosperm. The commercially available sweet corn cultivars derive their sugar and polysaccharide content through the genetic control of the recessive allele at the sugary-1 locus on chromosome 4. At optimum edible maturity, high quality sweet corn normally contains about 20 percent of its dry matter as sugar. Recently, Wann, et al[25] reported that total sugars could be increased markedly by using some of the endosperm mutants previously found by Creech;[26] namely, *ae* (*amylose extender* on chromosome 5), *du* (*dull* on chromosome 10), and *wx* (*waxy* on chromosome 9).

Their results are as follows:

Corn Selection	% Total Carbohydrates	% Starch	% Sugars
Golden Security (check)	75	32	22
M6212A (*ae wx*)	73	35	35
M6210B (*ae du wx*)	72	29	40

Thus, the new mutant genes were effective in raising the sugar levels in sweet corn to nearly double the amount found in the commercial sweet corn cultivar, Golden Security. The production of total carbohydrates and starch meanwhile were slightly reduced but essentially not affected insofar as the percentage of dry matter is concerned.

However, the starch content in the sugar corn lines with the mutant genes *ae wx* combination, when treated *in vitro* with pancreatic alpha amylase, was less digestible.[27] Therefore, from these rather preliminary tests we might state that the combination of endosperm mutant genes *ae wx* had the pleiotropic effect of doubling the sugars, while at the same time changing the starch structure to a form poorly digestible by the enzyme that normally digests starch in the small intestine. However, these results are preliminary and to my knowledge have not been verified.

One might ask whether low-digestible starch is an undesirable characteristic in the American diet, and an expert in human nutrition probably would pause for a great deal of thought on the matter.

BITTER PRINCIPLE IN CUCURBITS

Vegetables of the *Cucurbitaceae* family possess genes that control bitterness factors in their fruits, particularly cucumbers, squashes, and watermelons. Twelve bitter principals, known as cucurbitacins, have been found.[28] Plant breeders have no difficulty in selecting and breeding nonbitter fruits. However, these substances are of interest because they have been reported to be carcinostatic[29] as well as attractants for spotted cucumber beetles (*Diabrotica undecimpunctata* howardi Barber).[30]

Nonbitterness is apparently genetically controlled in watermelon by a complex system of a single recessive suppressor gene, *su bi*, that is active in the presence of the dominant gene for plant bitterness, *Bi*, along with a modifier gene, *Mo bi*, that modifies the amount of bitterness.[31]

Other undesirable compounds or unattractive characteristics are as follows:

Name of Compound or Characteristic	Mode of Inheritance	Authority
Pubescence Peaches	Single dominant factor between nectarines and peaches, heavy pubescence is dominant over light pubescence.	Blake and Connors[32]
Pit Astringency in Apricots	Probably multi-· factorial	Cullinan[33]
Antimetabolites	Not known but there are varietal differences	Unpublished data
Favism	Inheritance of the factor in *Vicia faba* L. is unknown. In susceptible humans the susceptibility to favism is apparently sex linked factor with incomplete dominance.	Zinkham, et al[34]

SUMMARY

Almost all papers on genetic research to remove undesirable compounds from vegetables and fruits have been published since 1970. Limited knowledge is available on the mode of transmission of genes in vegetables governing such traits as nitrate accumulation and formation of neuro-toxins and glycoalkaloids. Likewise, unattractive characteristics of vegetables and fruits such as bitter taste, excessive pubescence, production of flatulence and unattractive color change are apparently genetically controlled, but the gene action governing these conditions is not well understood. This paper reviews essentially all known research to date on genetic engineering to remove these undesirable compounds and unattractive characteristics.

References

1. Wolff, IA and Wasserman, AE: Nitrates, nitrites, and nitrosamines, *Science* **177** (4043):15-19, 1972.
2. Jackson, WA; Steel, JS; Boswell, VR: Nitrates in edible vegetables and vegetable products, *Proc Amer Soc Hort Sci* **90**:349-352, 1967.
3. Viets, FG and Hageman, RH: Factors affecting the accumulation of nitrate in soil, water and plants, *Agr Handbook* 413, USDA, Washington, DC, 1971.
4. Barker, AV; Peck, NH; MacDonald, GE: Nitrate accumulation in vegetables I. Spinach Grown in Upland Soils, *Agron J* **63**:126-129, 1971.
5. Maynard, DN and Barker, AV: Nitrate content of vegetable crops, *Hort Sci* **7** (3):224-226, 1972.
6. Cantliffe, DJ: Nitrate accumulation in spinach grown at different temperatures, *J Amer Soc Hort Sci* **97**:674-676, 1972.
7. Farrow, RP; Charbonneau, TE; Lao, NT: Research report on internal can corrosion. National Canners Assoc, Washington, D C, 1969.
8. Hageman, RH; Flesher, D; Gitter, A: Diurnal variation and other light effects influencing the activity of nitrate reductase and nitrogen metabolism in corn, *Crop Sci* **1**:201-204, 1961.
9. Schrader, Le et al: Nitrate reductase activity of maize hybrids and their parental inbred, *Crop Sci* **6**:169-173, 1966.
10. Liener, IE: Lathyrogens in foods, in *Toxicants Occurring Naturally in Foods*, Pub. 1354, Natl Acad Sci, Washington, D C, 1966, pp. 40-46.
11. Swaminathan, MS: Role of mutation breeding in a changing agriculture, in *Induced Mutations in Plants*. Int. Atomic Energy Agency, Vienna, 1969, pp. 719-733.
12. Schreiber, K: Naturally occurring plant resistance factors against the Colorado potato beetle (*Leptinotarsa decemlineata*) and their possible mode of action, *Zuchter* **27**:289-299, 1957.
13. Fontaine, TD et al: Isolation and partial characterization of crystalline tomatine, an antibiotic agent from the tomato plant, *Arch Biochem* **18**:467-475, 1948.
14. Kingsbury, JM: *Poisonous plants of the United States and Canada.* Englewood Cliffs, N J: Prentice Hall, Inc., 1964, p. 626.
15. New potato variety withdrawn. Press release (USDA 423-70), Feb 11, 1970.

16. Sanford, LL and Sinden SL: Inheritance of potato glycoalkaloids, *Amer Potato J* **49**:209-217, 1972.

17. Orgell, WH and Hibbs, ET: Human plasma cholinesterase inhibition *in vitro* by extracts from tuber-bearing *Solanum* species, *Proc Amer Soc Hort Sci* **83**:651-656, 1963.

18. Orgell, WH and Hibbs, ET: Cholinesterase inhibition *in vitro* by potato foliage extracts, *Amer Potato J* **40**:403-405, 1963.

19. Akeley, RV; Houghland, GVC; Schark, AE: Genetic differences in potato-tuber greening, *Amer Potato J* **39**:409-417, 1962.

20. Sinden, SL: Effect of light and mechanical injury on the glycoalkaloid content of greening-resistant potato tubers. (Abstract) 56th Annual Meeting of Potato Assoc Amer, 1972, pp 16-17.

21. Lovelace, FE; Liu, CH; McKay, CM: Age of animals in relation to the utilization of Ca and Mg in the presence of oxalates, *Arch Biochem* **27**:48-56, 1950.

22. Fassett, DW: Oxalates, in *Toxicants Occurring Naturally in Foods*. publ 1354, *Nat Acad Sci,* Washington, D.C, 1966, pp. 257-266.

23. Eheart, JF and Massey, PH, Jr: Factors affecting the oxalate content of spinach, *Agr Food Chem* **10**:325-327, 1962.

24. Murphy EL: The possible elimination of legume flatulence by genetic selection. Read before Protein Advisory Group Symposium on Nutritional Improvement of Food Legumes by Breeding, FAO, Rome, Italy, July 3-5, 1972.

25. Wann, EV; Brown, GB; Hills, WA: Genetic modification of sweet corn quality, *J Amer Soc Hort Sci* **96**:441-444, 1971.

26. Creech, RG: Carbohydrate synthesis in maize, *Advance Agron* **20**: 275-322, 1968.

27. Rosario, TL: Digestibility of starch granules from normal and mutant maize *(Zea mays L.)* kernels. PhD Thesis, Pennsylvania State University, 1971.

28. Rehm, S: Die Bitterstaffe der Cucurbitaceen, *Ergebn Biol* **25**:108-136, 1960.

29. Gitter, S et al: Studies on the anti-tumor effect of cucurbitacins, *Cancer Res* **21**:516-521, 1961.

30. Chambliss, OL and Jones, CM: Chemical and genetic basis for insect resistance in cucurbits, *Proc Amer Soc Hort Sci* **89**:394-405, 1966.

31. Chambliss, OL; Erickson, HT; Jones, CM: Genetic control of bitterness in watermelon fruits, *Proc Amer Soc Hort Sci* **93**:539-546, 1968.

32. Blake, MA and Connors, CH: Early results of peach breeding in New Jersey, bulletin 599, *N J Agr Exp Sta,* 1936.

33. Cullinan, FP: Improvement of stone fruits, *Yearbook of Agriculture,* USDA, 1937, pp 665-748.

34. Zinkham, WH; Lenhard, RE, Jr; Childs, B: Erythrocyte glutathione metabolism and drug sensitivity, *J Pharm and Exp Therapeutics* **122**: 85A, 1958.

The Economics of Genetic Engineering

Edwin A. Crosby, Ph.D

As we look at the factors which may influence the nutritional values of fresh fruits and vegetables and the prospects for using genetics to improve them, we cannot ignore the factor of economics which will dictate (1) whether our farms can produce fruits and vegetables of higher nutritional value; (2) whether our storage and marketing practices can be altered to protect these crops from loss of naturally occurring nutrients; and (3) whether the consumer can pay the price for better nutrition in the market place.

Let us look at why fruits and vegetables are included in the American diet. It seems there are two basic reasons, one nutritional and the other esthetic. From the standpoint of nutrition, principal consideration given to fresh fruits and vegetables should be in the area of vitamin content and, secondarily, mineral content. It is to be questioned, however, whether the average housewife thinks about nutrition when she buys these products or whether she isn't more concerned about variety, color, esthetics, flavor, texture and other factors in her menu planning. If we move to nutritional labeling, we may well have a large number of consumers awakened to the realization that fruits and vegetables are *not* generally good sources of nutrients from an economic viewpoint, and if consumers are truly concerned about insuring that their families receive enough vitamins and minerals, the easy, economical route to insure this end is to purchase them in pill form. It would indeed be difficult strictly on an economic basis to justify the purchase of fruits and vegetables for their nutritional values because the nutrients they contain can be better obtained, in terms of cost, from other sources. If tomorrow's housewife becomes a nutritional bug with a calculator in her pocket when she goes to the supermarket and thinks only of value in terms of nutrition, she will avoid the purchase of fruits and vegetables.

Let us explore some of the dollar and cents aspects of food production in terms of nutritional value. In view of the popularity and importance of tomatoes as a vegetable in this country, I would like to use this crop to illustrate certain points. The production of tomatoes for both the fresh market and processing in the United States in 1971 totaled 12,875,000,000 pounds. The total market value of the crop in raw form was $444,000,000. The USDA's *Agriculture Handbook No. 8, Composition of Foods* has been used as a standard reference for determining the average nutrients in the edible portion of food as purchased. *Handbook No. 8* reveals that in respect to vitamin C, the year-round average for whole, ripe, raw tomatoes is 102 mg per pound. From the 1970 Opinion Research Corporation Study of Public Attitudes Toward Added Vitamins in Foods (a study for Hoffmann-LaRoche), it is revealed that the Recommended Dietary Allowance (RDA) of vitamin C, 60 mg, could be provided for the fortification of foods at a cost of .0195 cents. For simplicity in calculations to illustrate the point I wish to make, I have rounded these figures to 100 mg per pound for vitamin C content of raw tomatoes and the price for an RDA of vitamin C to .02 cents. Using the foregoing figures, some interesting calculations can be made. For example, the total tomato crop produced in the United States in 1971 for the fresh and processing markets contained 21½ billion RDAs of vitamin C. In terms of the price of synthesized vitamin C, this was equivalent to $4,292,000 worth of this vitamin. In terms of the total market value of the crop, the value of the vitamin C was equivalent to slightly less than 1%. In other terms, each ton of tomatoes contained 66 cents worth of vitamin C, each pound 0.033 cents worth.

Let us assume for a moment that through genetic engineering we can increase the vitamin C content of tomatoes by 100% from 100 mg per pound to 200 mg per pound. Assuming the farmer could sell these tomatoes strictly on the true market value of the higher vitamin C content, based on the average yield of processing tomatoes in 1971, slightly over 21 tons per acre, this would be worth $13.86 per acre. For the fresh market grower it would be worth $4.42 per acre based on the average yield in 1971. Purely from the farmer's point of view in terms of the profitability of producing tomatoes in 1971, this would have been equivalent in value to slightly more than one-third ton of increased production per acre for processing or approximately 32 pounds for the fresh-market grower. In terms of value to Mrs. Consumer, strictly thinking economics, she could not afford to pay more than an additional 0.033 cents per pound for tomatoes with twice as much vitamin C as found in the present tomato varieties. In other words, if she is strictly interested in buying tomatoes for vitamin C content, this would be the fair increase in market value for tomatoes with twice as much vitamin C.

It is thus to be questioned whether there is economic justification for making vitamin C improvement in tomatoes a primary objective in a breeding program. If development of varieties with twice as much vitamin C is to be the main objective of a breeding program based on values above revealed, we have leeway for approximately one third of a ton shift in yield to offset a theoretical value increase due to the vitamin C. If the farmer is to produce tomatoes with higher vitamin C content, it must be done without any reduction in marketable yields. Furthermore, the consumer cannot afford to pay a higher price for tomatoes with twice the vitamin C content of present day varieties.

Using the same approach in evaluating vitamin A content of tomatoes, wherein the vitamin A value for whole, raw, ripe tomatoes, is taken as 4000 IU per pound and the dollar value of an RDA (5000 IU) is $0.003, the value of the vitamin A in the 1971 tomato crop was $3,090,000. This amounted to only 0.70% of the value of the total crop; or the vitamin A in a ton of processing tomatoes had a value of only 48 cents.

Using plant breeding to improve the vitamin A content in tomatoes has been proved by Purdue University to be very feasible. Several years ago Purdue introduced a tomato variety, Carored, having an average vitamin A content ten times that of conventional varieties. In spite of the significantly improved nutritional value of the variety in terms of vitamin A, the variety is of little commercial importance because of its orange-red color. At the present time, Purdue University is considering the introduction of a new variety having a higher red color than present varieties, but, interestingly enough, having approximately half the vitamin A content of most present day varieties.

It is evident that through genetic engineering, improved cultural techniques, and careful attention to storage and marketing practices, the nutritional qualities of fruits and vegetables can be improved. The real question, however, is whether improved nutritional value for fresh fruits and vegetables is truly important to the public welfare. Should this area of research be given priority or should we look to vitamin pills or fortified processed products as may be desirable to improve nutrition? What choices will the consumer make if we move to nutritional labeling of fresh fruits and vegetables as well as processed foods? If higher nutritional values in these products mean higher costs, the decision of the economy-minded buyer is obvious. A simple pill can well be a preferable alternative to paying for higher nutritional values in fresh fruits and vegetables. It is to be expected, however, that greater fortification of processed foods is forthcoming. For example, the fortification of tomato juice and possibly other tomato products with vitamin C such that an average serving will carry the RDA for this vitamin has been under consideration for some time.

As we move further into the new era of consumer concern for the quality of food, it is important that as scientists we not lose sight of reality and priorities in our role of providing for an adequate and varied food supply for the increasing American population. Do we have serious nutritional deficiencies in our fruits and vegetables and can we or should we do something about it? Will the consumer pay for improved nutritional values in this portion of the family's diet or will she seek a less expensive route to achieve nutritional balance? Do we really eat fruits and vegetables because of their *great* nutritional benefits or do we simply like them and want them as a part of our diet? Certainly there are certain fruits and vegetables such as citrus,

carrots, spinach, etc. which provide high levels of certain vitamins or minerals that are of importance in our diet. Since
alternatives exist in respect to what a particular woman may
feed her family, why not encourage selection from existing
good sources of nutrients rather than trying to improve nutrient
levels in present varieties? Why increase vitamin A in tomatoes
when carrots are available?

It is important that we, as scientists and administrators concerned about the future food supply of this country, do not
lose sight of practical realities in the political climate in which
we live. We should be leading and not following those who are
politically or emotionally motivated. Statements have been
made suggesting that in the movement toward modern, mechanized agriculture, we lost nutritional values that previously
existed in fresh fruits and vegetables. There are no substantive
data to establish this as fact. Rather, there are data to support
the conclusion that, in spite of the changes in varieties and
cultural systems that have been made in recent years, the
nutritive value of our products has not changed significantly in
the past thirty years. I say this partly on the basis of our own
Association's research in re-evaluating data published in *Agriculture Handbook No. 8* and obtained in 1941-43. In our update
of two major crops, tomato juice and whole kernel corn, we can
find no evidence that there is or has been a pronounced change
in the nutritive value of these products over the past thirty
years.

With affluence comes concern for many issues and certainly
not the least of these is concern for the quality of our food
supply. In spite of the fact that this concern may be justified in
some product categories, the overall nutritional content of our
food supply in 1969 far exceeded the nutritional requirements
of our population.

It is important to recognize priorities in meeting the need for
total food production. Priorities in respect to the production of
fruits and vegetables must include the factors which keep our
agriculture competitive in domestic and world markets. It is
only logical that emphasis be placed on factors such as yield;
harvestability; storage, shipping and handling qualities; color,

flavor, and general consumer acceptance; disease and insect resistance. When it comes to the economics of genetic engineering, these factors have been given preference in the past and
should continue to be given emphasis in the future. The battle
for survival in agriculture will continue and it will not be helped
by a redirection of emphasis in breeding and development of
new varieties toward nutritional quality improvement. Certainly
we should not ignore genetic manipulation and its influence on
major nutrients; the FDA will not allow it. We should be careful
not to lower levels of important nutrients even though we must
continue to stress economic factors relating to production,
marketing and utilization. A very simple illustration is the need
to concentrate on the development of new disease- and insect-
resistant varieties in accordance with the new regulations and
prohibitions that govern and restrict the use of organic pesticides.

It is hardly necessary to point out the time lag that would be
involved should we choose to upgrade the nutritional value of
apples or any of our major fruit crops. Any substantive change
in the varietal picture of tree fruits could hardly be accomplished in this century. Even in the shorter time frame of
vegetable breeding and development, we must not lose sight of
the fact that it isn't simply the introduction of a new variety of
higher nutritional value, it is the introduction of many new
varieties adapted to different uses and to the variable environ-
ments under which they will be produced commercially in this
country.

It is extremely important to the welfare of American agriculture that those responsible for the breeding and the development of new varieties of fruits and vegetables not be caught in
the nutrition stampede. Furthermore, if emphasis on nutrition
and nutritional improvement in the products consumed by
Americans is to receive increased attention, it is quite likely that
simple fortification of basic foods marketed in other than fresh
form will provide more practical solutions to identified problems.

In the study previously mentioned, conducted by the Opin-
ion Research Corporation, it was determined that the cost of
vitamin fortification for nine vitamins identified as important to

human health (100% of RDA or reasonable estimate of the Minimum Daily Requirements for three vitamins for which no MDR has been established) would be *only one quarter of a cent.* Thus the food processing industry could put all the vitamins necessary according to present day knowledge in a fortified food without influencing the true value of that product by more than a quarter of a cent. It hardly seems reasonable to direct our breeding efforts in fruits and vegetables toward this type of effort. As for the minerals, which we may look to vegetables and fruits to supply in some cases, these can also be provided at low cost through fortification.

In spite of the possibilities and promise of doing certain things through genetic engineering to improve the nutritional values of our fruits and vegetables, where are the economic reasons justifying such effort?

Vitamin Induction in Fruits
and
Vegetables with Bioregulators

V. P. Maier, Ph.D., and H. Yokoyama

There is a continuing need to improve the nutritional value of the human diet by making every food crop as nutritious as possible.[1] Part of that job involves preventing loss of nutritive value as the crop proceeds from the farm to the consumer. The other part of the job involves finding ways to improve the inherent nutritive value of the crop at maturity. One approach to the latter objective, which has received only minor attention, is that of specific induction and accumulation of desired substances through the action of applied bioregulatory agents either before or after harvest. The function of these bioregulators is to control portions of *specific* biosynthetic and/or metabolic pathways and thereby bring about an accumulation of desired constituents. While this approach can be applied to either the accumulation or removal of plant constituents,[2] this discussion will be limited to the subject of the accumulation of vitamins.

While it will not be a simple task to achieve vitamin accumulation with bioregulators, the biochemical principles upon which this approach rests are established and much of the specific information needed about the biosynthetic pathways and composition of individual plants is already available.

The genes determine whether a plant has the capacity to synthesize a given compound. However, the interaction of the genes with complex regulatory mechanisms, which respond to stimuli from the environment, determine the amount, if any, of the compound made and accumulated in the various tissues of the plant.[3] Consequently, demonstration of even trace amounts of a vitamin in any tissue of a plant confirms that the necessary genes are present for biosynthesis of the vitamin by the plant. Once the presence of the gene is established, studies aimed at discovering appropriate stimuli with which to cause the regulatory system to initiate or increase biosynthesis and accumulation of the vitamin are then needed. In practice the situation will undoubtedly be more restricted than the general case stated above, in that certain vitamins and plants will be inherently easier to control than others.

Of the two general classes of stimuli, chemical and physical, chemical stimuli are potentially the most specific and the most numerous. This paper will discuss only chemical stimuli and they will be herein denoted by the term "bioregulators".

The very difficult task of finding bioregulators will be aided somewhat by employing comparative biochemical techniques. By studying other appropriate organisms that accumulate appreciable amounts of the desired vitamin, the needed bioregulator might be isolated and identified. Studies on the vitamin-deficient plant will then permit evaluation of the effectiveness of the bioregulator and its analogs. In theory, once the appropriate regulator substances are found they will probably be effective on a variety of crops.

This approach to production of crops having enhanced nutritional value should complement other methods aimed at the same goal. In particular, however, this approach appears to be advantageous for use on tree crops. Because of the slowness and great expense of replacing bearing trees or of topworking them to a new variety, the breeding approach to vitamin enhancement of tree crops can be slow and expensive. In addition, it may not always be possible to breed into a single variety all of the desirable traits wanted.

An example of the successful use of this approach is the enhancement of the provitamin A content of citrus fruits. Treatment of mature citrus fruits by immersing them for 30 seconds in a solution containing 5000 ppm of a bioregulator of carotenoid biosynthesis caused a sharp increase in net carotenoid synthesis and an accumulation of lycopene and the provitamin A, gamma carotene. This happens whether the fruit is treated pre- or post-harvest and without affecting the other provitamin A's present in the fruit at the time of treatment.[4] Thus, the provitamin A content of tree-ripe navel oranges was increased from 100 I.U./100 gm (dry weight peel) to 1500 I.U./100 gm, and of tree-ripe grapefruit from about 20 I.U./100 gm (dry weight peel) to 1900 I.U./100 gm.[5] At the same time there was a large net synthesis of lycopene and its carotenoid precursors, but no change in the terpenoid flavoring constituents.

Studies with the carotenogenic mold *Blakeslea trispora* have shown that lower concentrations of the bioregulator favor gamma carotene accumulation over that of lycopene, while at the same time greatly stimulating net carotenoid biosynthesis.

Other data[6] indicate that the bioregulator stimulates net carotenoid synthesis by increasing the rate of enzyme synthesis, possibly through gene derepression. On the other hand, accumulation of gamma carotene and lycopene appears to be caused by inhibition of the activity of specific cyclases of the pathway.

A number of active analogs of the discovery bioregulator, 2-(4-chlorophenylthio)-triethylamine hydrochloride (CPTA), have been synthesized and work is in progress on the natural citrus bioregulator.

The ability of the bioregulator, CPTA, to stimulate and direct carotenoid biosynthesis has been demonstrated in a wide variety of higher and lower plant tissues including roots of carrot and sweet potato, fruits of apricot, prune, peach, the mycelia of *Blakeslea trispora* and *Phycomyces blakesleeanus* and cells of the photosynthetic bacterium *Rhodospirillum rubrum.*[7]

References

1. Senti, FR: Address before the American Phytopathological Society, Philadelphia, Pa., August 17, 1971.
2. Maier, VP: *Proceedings First International Citrus Symposium,* Riverside: University of California Publ Dept, 1969, vol 1, p 235.
3. Davis, D: *Science* **170**:1279, 1970.
4. Yokoyama, H; Coggins, CW, Jr; Henning, GL: *Phytochemistry* **10**: 1831, 1971.

5. Yokoyama, H; DeBenedict, C; Coggins, CW, Jr; Henning, GL: *Phyto-chemistry* **11**:1721, 1972.
6. Hsu, WJ; Yokoyama, H; Coggins, CW, Jr: *Phytochemistry* **11**:2985, 1972.
7. Coggins, CW, Jr; Henning, GL; Yokoyama, H: *Science* **168**:1589, 1970.

Appendix

TASK FORCE RECOMMENDATIONS
*SYMPOSIUM ON NUTRITIONAL QUALITIES OF
FRESH FRUITS AND VEGETABLES

I. Research Centers

Interdisciplinary research centers concerned with nutrition and food quality should be established within the Agricultural Research Service of the U.S. Department of Agriculture. Five or six such centers are needed, each including facilities and staff for research on fruit and vegetable breeding, production practices, harvesting, processing, storage, and transportation. Laboratories and other facilities for evaluation of the nutritional qualities of foods, using experimental animals and human volunteer subjects, should also be included in each of the centers. The staffs of these centers should comprise all of the scientific disciplines required for research on all aspects of the problem from production of food crops to utilization of the nutrients when these crops are consumed by people.

In order to maximize use of existing scientific facilities and to encourage graduate training in the scientific disciplines involved, the centers should be located at state land-grant universities where possible. In some instances these centers can be housed in existing federal and state buildings; thus the cost of this program may be very small in comparison with the potential benefits.

---

* Sponsored by the American Medical Association's Council on Foods and Nutrition, in cooperation with the Agricultural Research Service, USDA, November 9-10, 1972, San Diego, California.

II. **Agricultural Production Practices**

A. *Effects of fertilization, irrigation and other production practices upon nutritional quality and concentration of toxic substances.*

Research on the effects of production practices upon concentrations of essential minerals in the edible parts of fruits and vegetables has not been adequate. Most of the exisitng data on the effects of fertilizers, for example, on mineral concentrations in plants are concerned with mineral concentrations in leaves, with relatively little data on seeds and fruits. In addition, the effect of fertilization or other cultural practices upon the concentration of alkaloids or other toxic materials in fruits and vegetables needs to be evaluated. Studies of forage crops indicate that the concentration of certain alkaloids is increased by high applications of nitrogen fertilizers.

B. *Control of the levels of nitrate in fruits and vegetables.*

Measurements of the nitrate levels in various crops at various stages of growth and correlation of these measurements with yield and with nitrate levels at harvest are needed in order to provide a basis for production systems that will foster optimum yields with minimum nitrate concentrations. The availability of the nitrate electrode for measuring nitrate concentrations in plant tissue now makes it possible to monitor nitrate levels in field-grown plants throughout the growing season.

C. *Effects of location of production upon nutritional quality.*

It has been established that plants produced in certain locations are consistently higher in iodine content than plants produced in other locations, and that these regional differences can be related to needs in human nutrition for iodine supplementation. A regional pattern of selenium concentration in plants has also been described, but effects of this pattern upon human nutrition have not been established. There is a need to investigate the possibility of regional patterns in the concentration of other nutrients, especially trace minerals and magnesium, in fruits and vegetables.

III. **The availability of nutrients in fruits and vegetables.**

The value of fruits and vegetables as sources of essential

minerals is presently based upon the total concentrations of minerals in the edible portions rather than upon evidence that the minerals are available for utilization by the human body. There is evidence of fairly low availability of some minerals, especially zinc, to experimental animals. A major program to evaluate the availability for human nutrition of minerals and other nutrients in fruits and vegetables could provide a basis for improvement of the nutritional status of people in the U.S. as well as in developing countries.

IV. Effects of waste disposal practices upon nutritional quality.

Many centers of fruit and vegetable production are located near centers of population where recycling of wastes by applying these wastes to land is being considered. Research is needed to determine whether or not hazardous accumulation of toxic elements such as cadmium and mercury may result from this process. Also, the possibility of transmitting communicable diseases through application of various types of wastes must be thoroughly studied.

V. Plant breeding and genetic engineering
A. Identification of nutrients and toxicants which are important in breeding programs.

If progress is to be made in the breeding of fruits and vegetables of superior nutritional quality, it is necessary to identify the most important nutrients to increase, or the most important toxic materials to decrease. For example, there has been relatively little work on factors affecting folic acid concentration in plants. Are nutritionally important increases in folic acid concentration in vegetables possible through plant breeding programs? Breeders of food crops must be aware of the possibility of inadvertent increase of toxic materials arising from breeding plants with increased resistance to insects and diseases. Research directed toward identifying plant resistance mechanisms is important in identifying possible toxicants. Establishment of the interdisciplinary centers previously described (see number I would increase contacts between plant breeders and other scientists and facilitate identification of the best objectives for plant breeding programs.

B. *Genetic control of biosynthetic pathways.*

There is little understanding of biosynthetic changes in higher plants that are associated with the accumulation and/or decrease of compounds important in human nutrition. The identification of intermediate products, the nature of inheritance and the comparative biochemical similarities between species would provide efficiency in effecting significant changes via plant breeding.

C. *Development of analytical methodology.*

If plant breeders are to work toward enhanced nutrient concentrations in fruits and vegetables, methods suitable for measuring these concentrations in large numbers of plants must be available. For example, it is nearly impossible to work on improvement of vitamin E levels in plants at present due to the time required for meaningful determination of this vitamin in plants. If rapid, accurate methods for measuring the concentration of any essential nutrient in fruits and vegetables are developed, progress toward improvement of the plants in this nutrient is very likely. An excellent example of this effect is the use of dye binding capacity of cereal protein by plant breeders working toward increased concentrations of lysine in cereals.

D. *Plasticity of fruit and vegetable germ plasm.*

The total range of properties inherent in fruit and vegetable species needs to be carefully evaluated. Heritable characteristics that are of value in the storing and handling of fruits and vegetables need to be identified along with those characteristics, such as disease-resistance, that are of primary value in crop production. Germ plasm banks may contain genes that are especially useful in the preservation of nutritional quality in harvested crops.

E. *Varietal responses to harvest and handling practices.*

The effects of mechanical harvesting and various storage and handling practices upon different varieties of important fruit and vegetable crops should be evaluated, and varieties well-suited to specific combinations of production, harvest, handling and storage should be developed. An example of a specific project in this area would be development of leafy vegetable varieties characterized by very slow conversion of nitrate to nitrite during postharvest processes.

F. *Increases in protein quantity and quality.*

Although the Symposium's primary concern was with the vitamin and mineral values of fresh fruits and vegetables, these foods are also important sources of protein in human diets, and are likely to become more important for this purpose as population pressures upon food supplies increase. Worthwhile advances in improvement of the amino acid balance in fruit and vegetable proteins appear to be possible and research toward this objective should be encouraged.

VI. Post-harvest handling of fruits and vegetables

A. *Time-temperature-environment tolerance of fruits and vegetables.*

The effects of various temperatures and various types of environment, over periods of time, upon different nutrients in different fruits and vegetables should be determined in a systematic fashion. At present, post-harvest handling procedures must be based upon fragmented evidence, with visible signs of decomposition being used as the major index for establishing handling procedures. Basic data on the effect of specific storage conditions for varying periods of time upon specific nutrients in fruits and vegetables will permit a more rational design of transportation and storage systems.

B. *Interaction of plant materials with various organisms.*

Much of the post-harvest deterioration of fruits and vegetables is due to the attack of microorganisms. The conditions favoring or inhibiting the growth of specific microorganisms on fruits and vegetables need to be defined, and the biochemistry of microorganism-induced changes requires study.

C. *Methods for predicting "shelf life" of fruits and vegetables.*

Brokers and grocery store buyers, as well as the ultimate purchasers of fruits and vegetables, must use outward appearance of the produce to estimate "shelf life". The chemical processes that result in termination of shelf life probably are well underway before evidence of decomposition is visible. If quick tests for the progress of these processes could be developed, they could provide the basis for design of improved handling, storage and marketing systems for fruits and vegetables, and increase consumer acceptance of these materials.

D. *Nutritional effects of semi-processing.*

Many fruits are treated with waxes, dips, coloring agents, etc. during post-harvest processes. Many of these treatments are designed to improve the appearance of the fruit. The effect of these treatments upon the nutrients in the fruit should be investigated.